HOW THE WORLD WORKS
PLANET EARTH

From molten rock in space to the place we live

Anne Rooney

Picture Credits

Alamy Stock Photo
36 Nick Upton, 41 The Natural History Museum, 54 Chris Hellier

Getty Images
11 Stocktrek Images, 35 Ullstein Bild dtl, 90 Photos.com, 104 Jack Goldfarb, 194 Mangiwau

NASA
140, 57, 52 NASA/JPL, 46 NASA/Simone Marchi

Public Domain
8, 10, 112, 153, 163, 182

Science Photo Library
13 Pekka Parviainen, 15 Oxford Science Archive/Heritage
European Southern Observatory, 25 Walter Myers, 48 Wa
Henning Dahloff, 61 Henning Dahloff, 63 Oxford Science
Archives/American Institute of Physics, 92 Mikkel Juul Jer
London, 105 B. Murton, Southampton Oceanography Ce
Tim Brown, 111 Gary Hincks, 113 Spencer Sutton, 116 B
P Rona / OAR / National Undersea Research Program / N
126 Chris Butler, 128 Dr Juerg Alean, 130 Gunilla Elam,
Bizley, 154 Microgen Images, 156 Science Source, 160 Pa
170 Paul D Stewart, 177 Joe Tucciarone, 181 Joe Tucciarc
Mikkel Juul Jensen, 191 John Sibbick, 193 NASA, 199 Ma
Foundation, 205 John Sibbick

Shutterstock
2, 6, 26, 29, 30, 34, 40, 55, 56, 66, 69, 70, 72, 73, 73, 74, 77, 78, 79, 80, 81, 83, 83, 84, 84, 86, 87, 88, 99, 99, 100, 100, 101, 103, 106, 108, 109, 114, 122, 123, 124, 125, 131, 135, 139, 143, 144, 144, 146, 148, 149, 151, 154, 158, 161, 166, 168, 170, 171, 173, 175, 176, 178, 179, 181, 185, 187, 188, 190, 192, 195, 196, 200, 203, 14

Wikimedia Commons
16, 21, 24, 28, 30, 32, 40, 43, 44, 57, 62, 63, 64, 65, 75, 76, 82, 85, 86, 98, 105, 107, 122, 127, 128, 132, 133, 134, 152, 155, 157, 159, 161, 164, 169, 172, 180, 184, 202

ARCTURUS

This edition published in 2020 by Arcturus Publishing Limited
26/27 Bickels Yard, 151–153 Bermondsey Street,
London SE1 3HA

ISBN: 978-1-83857-378-2
AD007330UK

Printed in Singapore

Contents

Introduction
MIRACULOUS EARTH

'In every out-thrust headland, in every curving beach, in every grain of sand there is a story of the Earth.'
Rachel Carson, 'Our Ever-Changing Shore', 1958

Stories concerning the state of our planet – changes in its climate, ecosystems and atmosphere – are in the news every day. Over 4.5 billion years of existence, Earth has transformed from a hostile, sterile lump of hot rock whirling in space to a temperate planet of water and soil, cloaked with greenery and teeming with life.

The rocky coastline of Rinca in Indonesia outlines near-circular bays that reveal the island's history of ancient volcanism. The ocean now floods prehistoric volcanic craters.

Humans have studied the history of Earth for hundreds of years. Sometimes held back by supernatural beliefs and misinterpretations, we have now come to a good understanding of Earth's past and how it behaves. The challenge today is how we can put that knowledge to good use and keep the planet habitable for all life upon it.

Adapt or die

Life on Earth has always survived by adapting to changing conditions. In turn, life has also affected changes upon its environment. Sometimes the changing conditions have resulted in dominant life forms dying out and others coming along to take their place. Temperatures and sea levels have risen and fallen, mountains have grown and crumbled, seas have opened up and lands closed back together again.

The story of Earth is written in the rocks beneath our feet. But until we went

searching for it, the story remained hidden. Humankind has taken hundreds of years to read a a part of it, and there is still a lot to uncover and learn. We are the first beings on Earth to know and understand what preceded us, yet in geological time we have been here for only the blink of an eye.

Modern humans evolved just 200,000 years ago. If we think of the history of Earth as a clock and the present moment as midnight, then humans appeared a few seconds ago. If we imagine the clock as a single year, then vertebrates emerged on 20 November, mammals on 13 December, and modern humans arrived at 23:36 on 31 December. Agriculture began at 23:59 the same day, and the Industrial Revolution started a little after two seconds to midnight. Who knows what the New Year will bring?

A geological clock showing events from the formation of Earth over 4.5 billion years ago to the evolution of humans.

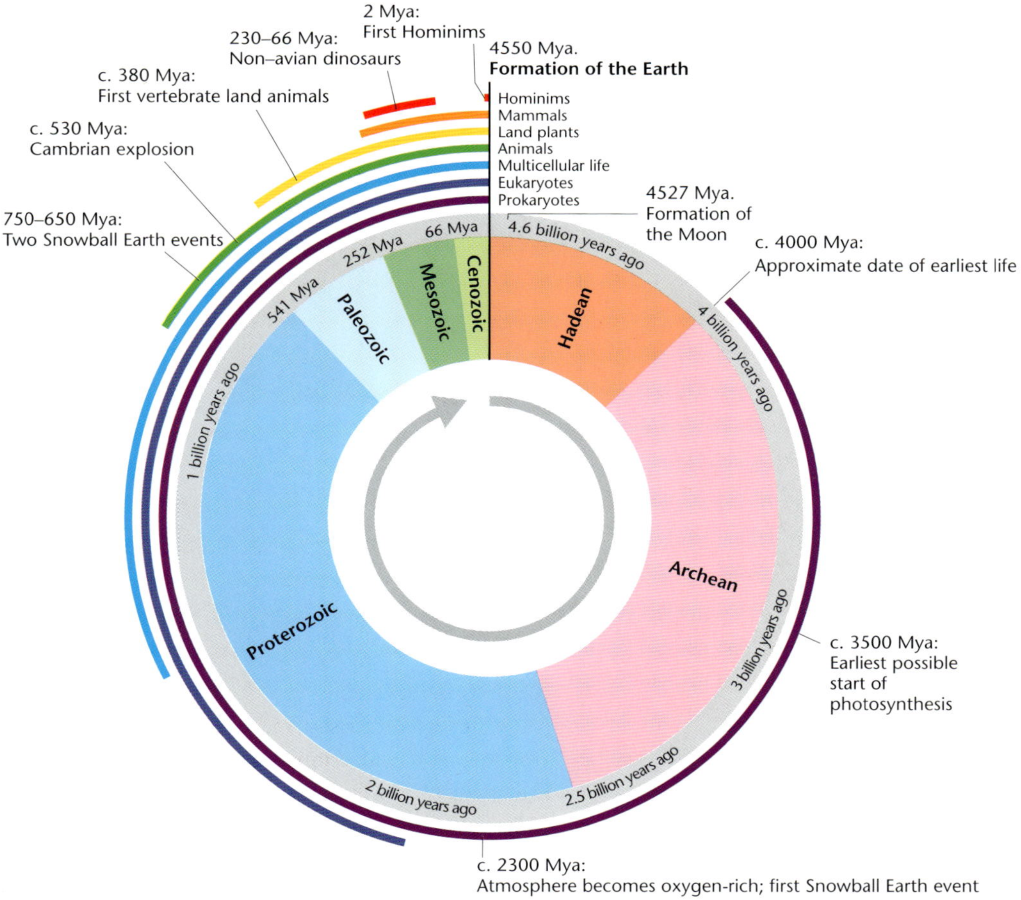

2 Mya:
First Hominims
230–66 Mya:
Non–avian dinosaurs
4550 Mya.
Formation of the Earth
c. 380 Mya:
First vertebrate land animals
Hominims
Mammals
Land plants
Animals
Multicellular life
Eukaryotes
Prokaryotes
c. 530 Mya:
Cambrian explosion
4527 Mya.
Formation of the Moon
750–650 Mya:
Two Snowball Earth events
c. 4000 Mya:
Approximate date of earliest life
66 Mya
252 Mya
541 Mya
4.6 billion years ago
4 billion years ago
1 billion years ago
3 billion years ago
2 billion years ago
2.5 billion years ago
Cenozoic
Mesozoic
Paleozoic
Hadean
Archean
Proterozoic
c. 3500 Mya:
Earliest possible start of photosynthesis
c. 2300 Mya:
Atmosphere becomes oxygen-rich; first Snowball Earth event

THE BRIGHTER BOROUGH
Wandsworth

9030 00007 6596 0

CHAPTER 1

Out of **CHAOS**

'In the Beginning how the Heav'ns and Earth
Rose out of Chaos'
John Milton, *Paradise Lost*, Book 1, 1674

A whirling cloud of gas and dust was the cradle of our planet and its companions. The story of Earth begins with the scattered particles of hundreds of destroyed stars, moulded by the gravity and heat of the forming Sun. But we need to step back even further in time than the 4.57 billion years during which the solar system has existed to see where the raw materials for world-building came from.

An artist's concept of a rotating disk of gas and dust around a forming star.

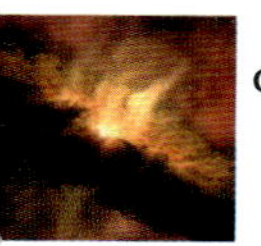

Making matter

Countless generations of stars lived and died before Earth's star, the Sun, was formed, and in doing so they produced the ingredients to make all the planets of the solar system. Every atom in our planet, and in our bodies, was forged in a long-dead star or in its cataclysmic end. Our connection with the universe is intimate and eternal: we humans, and everything around us, are stardust.

The nuclei of hydrogen atoms formed in the first second of the universe. A few minutes later, some of them got together to make the nuclei of helium, deuterium (heavy hydrogen) and lithium. They couldn't capture electrons and form atoms for another 380,000 years, as they had to wait for the universe to cool considerably. When it did, the atoms of hydrogen and helium that formed would be the raw material for future stars.

Starry ancestors

The first stars lit up only around 100 million years after the start of the universe. There have been many generations since then, but they have all worked in much the same way. Each star fuses hydrogen into helium for most of its life, releasing energy in the process. It is said to be on the 'main sequence' while doing this. When at last a star runs out of hydrogen it begins to fuse helium, making carbon. Then carbon is fused to make oxygen and other elements, and so on up until it makes iron (atomic number 26), the heaviest element that can be made in the heart of a star.

If the star is relatively small, its life ends there. Its outer layers blow away into space, leaving a white-hot core of iron which will cool over trillions of years. The life of a large star, however, ends with a very spectacular event called a supernova. The star collapses in on

Above: *A false-colour image of the remnant of Cassiopeia A, the youngest supernova in our galaxy. Cassiopeia A forms a cloud of gas and dust extending light years into space, carrying the material produced in the star's life and death.*

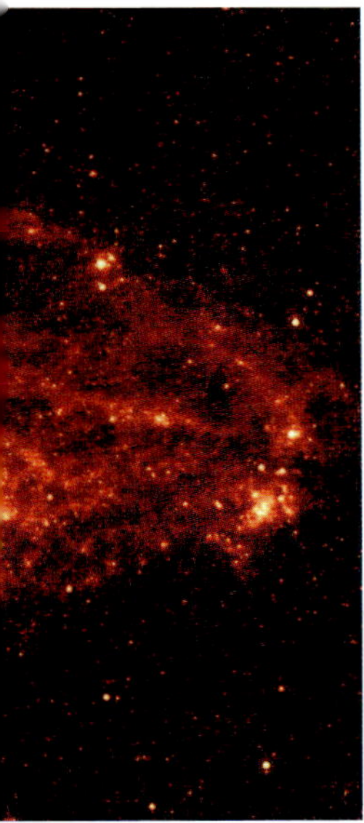

Left: *Vast clouds of cosmic dust in the Andromeda galaxy are revealed by an infrared telescope. From this dust, new stars and planets can form.*

itself under immense gravity and then bounces off its own nucleus, blasting its material outwards to be scattered into space. The huge release of energy is enough to force even iron atoms together, forming chemical elements right up to the heaviest naturally occurring element, uranium (atomic number 92). The entire mix of elements, exotic and mundane, is hurled out into the interstellar medium (the mix of dust and gas that is spread through space).

By the time that the Sun was forming, the interstellar medium was a rich cocktail of all the naturally occurring elements created in the hearts of stars and in their catastrophic demise.

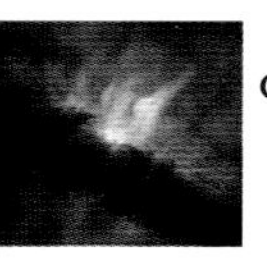

Clouds on the horizon

The gas and dust from all these previous stars, and the remnants of the primordial gases from the origins of the universe, are not distributed evenly through space. Gravity acts to draw matter together – even matter as tiny as atoms of gas. Wherever, by chance, there is a higher density of matter, more is drawn to it. Throughout space, vast molecular clouds of gas and dust hang suspended, poised in an equilibrium that prevents them either dispersing or contracting. If something disturbs that equilibrium, star-building can begin.

Seeding a solar system

Nearly 4.6 billion years ago, something – perhaps a passing star or the shockwaves from a nearby supernova – disturbed the molecular cloud of dust and gas from which the solar system formed. At first, pockets of dust and gas accumulated, making areas of greater density than elsewhere in the cloud. As each area gained in mass, so its gravity increased and it attracted ever more matter. Each centre of mass was on the way to becoming a star. One of them was the Sun.

The process escalated, with matter drawn closer and closer together so that it was under increasing pressure and became ever hotter. With most of the mass in the middle, the system began to rotate.

A forming star attracts matter, most of it falling into its rotating central body. Although at least 99.8 per cent of the mass from our solar nebular was drawn into the Sun, a tiny proportion was left over, whirling around it in space. During the course of about 100,000 years, what started as a vast cloud of gas and dust resolved into a thin disk around the central mass – much as spinning a ball of pizza dough transforms into a flat pizza base.

Increasing pressure pushed up the temperature until the ball of gas was so hot it began to glow, becoming a proto-star. And when a proto-star reaches a critical mass, it collapses inwards, triggering nuclear fusion. The pressure at the heart of the star is so great that it forces hydrogen atoms together, and the fusion of helium begins. All stars, including our own Sun, form like this. Within about 50 million years of the start of the collapsing cloud, the Sun burst into life as a main sequence star, blasting heat and light out into space.

Collapsing clouds

Although this discovery sounds quite modern, something like it was first suggested in 1734 by the Swedish scientist and theologian Emanuel Swedenborg. He proposed that the Sun was surrounded by magnetic particles of a coarser nature than the Sun itself and these rotated at the same rate as the 'solar vortex' in which he thought the Sun exists and from which it draws its energy. Through some kind of compression, the particles became 'grosser' and formed a shell over the surface of the Sun. Over time, this moved further away, becoming a 'belt or broad circle' surrounding the Sun. As it moved outwards, the belt was stretched to breaking point. The larger chunks became planets and the smaller chunks fell back inwards, becoming sunspots (which at the time were thought to be bodies moving across the face of the Sun).

Swedenborg experienced an intense spiritual awakening which led him to give

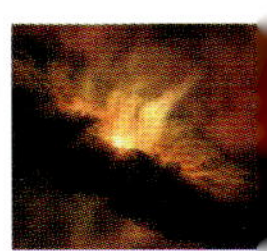

up science, so he never took his model further, but the German philosopher and scientist Immanuel Kant developed it into a nebular hypothesis in 1755. In his *Universal Natural History and Theory of the Heavens*, Kant described nebulae, or clouds of gas,

Below: *'Like trying to build a skyscraper in the middle of a tornado': how astronomer Henry Throop described the attempt to form new stars inside protoplanetary disks. The dust clouds shown here are in the Orion nebula.*

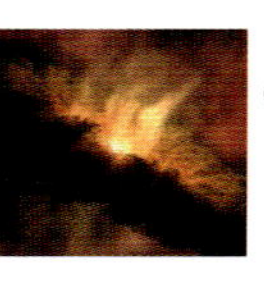

rotating, collapsing and flattening under the influence of gravity, finally forming stars and planets in a sequence very like the modern model. He suggested that the nebulae visible at his time with a telescope were regions of active star-building – which was indeed the case.

The French nobleman Pierre-Simon Laplace arrived independently at a more detailed formulation of a nebular hypothesis in 1796. He argued that the Sun was originally a hot, gassy cloud that extended over the full volume of the current solar system, then cooled and contracted to form a proto-solar nebula. As this spun and flattened it shed rings of gases from which the planets condensed.

ANGULAR MOMENTUM

The angular momentum of a system is given by the formula *mvR*, where *m* is the mass of an object moving in a circular orbit with radius *R* at velocity *v*. Angular velocity can never be destroyed, though it can be transferred in a system. In Laplace's model, all the angular momentum of the solar system would have originally been present in the nebular disk, and most of it would now be concentrated in the Sun. For this to work, the Sun would have to be rotating much more quickly than it is. (It takes about 25 days to rotate.) In fact, most of the angular momentum of the solar system is in Jupiter, Saturn, Uranus and Neptune.

However, if Laplace's account were accurate, the planets would be orbiting the Sun more slowly than they are. The Sun has the vast majority of the solar system's mass, but only 1 per cent of its angular momentum. The problem with angular momentum was demonstrated by the astronomer Forest Moulton in 1900, and the nebular theory fell out of favour for most of the 20th century.

In 1905, Moulton and the geologist Thomas Chamberlin proposed a theory to replace Laplace's hypothesis. They suggested that a wandering star approached the Sun closely enough to draw material out in spiral arms, ejecting material from it. After the star had passed, material left spinning around the Sun condensed, some of it forming small planetesimals, some larger protoplanets. These collided and combined over time, building up the planets and their moons. The leftover debris became asteroids and comets. Although most of this theory has been overtaken by more recent discoveries, the idea of planetesimals has survived into the modern account of the solar system's genesis.

The current form of the nebular theory is called the Solar Nebular Disk Model (SNDM) and originated in the work of the Soviet astronomers Viktor Safronov and Evgenia Ruskol. Safronov's *Evolution of the protoplanetary cloud and formation of the Earth and the planets* was published in 1969 and translated into English in 1972. Safronov and Ruskol realized that the speed of planetesimals in the disk was constantly changing as they came near to others and their gravitational fields interacted to slow them down or speed them up. When

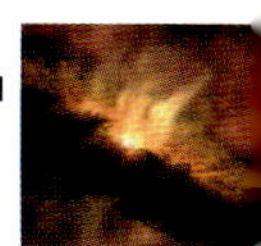

THE SUN'S EVIL TWIN

Several astronomers have suggested that the Sun was originally half of a twin-star system. Clearly, the twin has now gone. In the early 20th century, the American astronomer Henry Norris Russell said that a passing star had struck the twin, throwing out enough debris to form the planets. In Britain, Raymond Lyttelton thought the twin had been wrenched away by the intruding star, dropping enough material to form the planets, while Fred Hoyle suggested the twin went supernova and shed its material in the Sun's vicinity. The Dutch astronomer Gerard Kuiper argued that the proto-twin had never actually formed into a star but became the planets instead.

The English astronomer James Jeans suggested in the 1920s that a massive star closely approaching the young Sun drew out a cigar-shaped 'filament' of material which eventually broke off. The material cooled and condensed into the planets. The arrangement of the planets with the largest, Jupiter, in the middle of the system, reflects the shape of the filament.

collisions occurred, the outcome would depend on the speed of the colliding bodies. If they moved so fast that any broken bits achieved escape velocity, those chunks would be lost. If they moved slowly, the chunks would be pulled back in and the planetesimal would accrete more mass over repeated collisions.

Around the same time, Canadian astrophysicist Alastair Cameron was working on the distribution of radioactive isotopes and what this reveals about the development of the solar system. In 1975, he gave a lecture outlining the evolution of the solar system from the formation of the Sun through the collapse of a cloud of gas and dust, the subsequent formation of the protoplanetary disk, and growth of the gas and rocky planets from it. Cameron developed computer models using Safronov's data which led repeatedly to a similar arrangement of inner planets. They show Earth and the other rocky planets growing from a series of collisions between large protoplanets hurtling around the Sun

at speed. This model is at the heart of current theory.

Work at the start of the 21st century has refined the model and added some timings. It's now thought that the dust and gas of the nebular disk condensed over a period of just two to three million years.

From disk to planets

So far we have only seen rather vague suggestions about planets condensing from a cloud of gas and dust. Safronov's work was important because, at the time he was writing, there was little solid work on planetary formation. There were two broad approaches: the planets formed from the same material and at the same time as the Sun, or they formed separately and were captured by the Sun. Safronov, working in an under-resourced department in the Soviet Union during the Cold War, was of necessity a theoretician. His American contemporaries were using telescopic observations of comets and asteroids and studying the chemistry of meteorites in their attempts to unravel the making of Earth and the other planets.

Relying only on mathematics, Safronov began with the premise that the planets formed from an elliptical disk of dust, ice and gas all orbiting the Sun in the same direction. Earth and the other rocky planets grew from small particles that clumped together, their combined gravity then attracting more and more material. As a clump grows, it has greater gravitational attraction, so collects more dust, and grows further. A larger clump would also have had more encounters than a small one. It would smash apart some particles or knock them off course; others it would drag into itself. Working on the nature of collisions between particles, Safronov discovered that while fast collisions would result in lumps bouncing off one another, disrupting

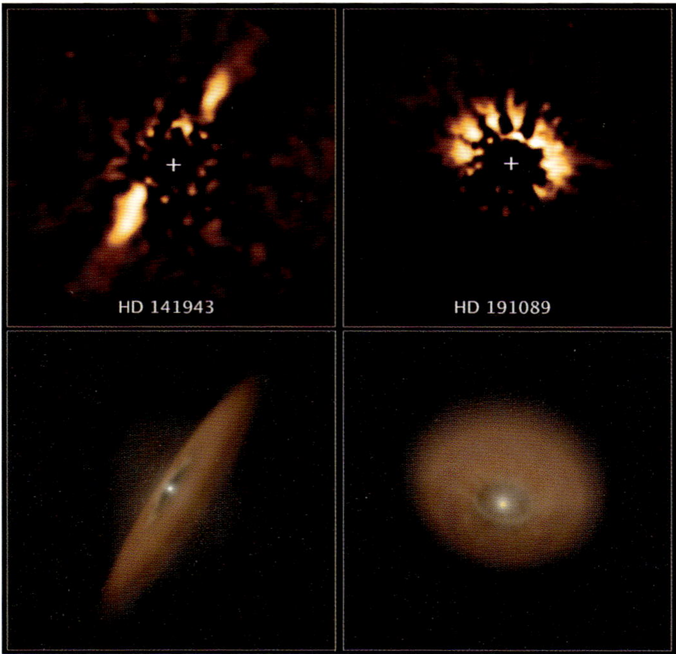

Dust disks around forming stars, seen edge on (left) and from above or below (right). The top images are infrared photos taken by the Hubble Space Telescope; the lower ones are visualizations drawn from the photos.

HOW TO MAKE A PLANET	
Stage 1: Dust settles from the nebula into a disk	Thousands of years
Stage 2: Dust and gas form clumps up to 1 km across	< 1 million years
Stage 3: Runaway growth up to 1,000 km (621 miles) across	A few hundred thousand years
Stage 4: Planet embryos grow through collisions	10–50 million years

their paths or breaking apart, lower energy collisions would lead to particles sticking together in clumps. Slowly, some of these clumps grow into planetesimals, sweeping up matter moving in their orbit.

Research in 2014 suggests that the cores of rocky planets began to form just 100,000–300,000 years after the start of the solar system. Then for up to 50 million years these embryonic planets crashed into one another and reconfigured to give the current terrestrial planets. It's impossible to pinpoint when Earth became Earth, as this stage of its construction involved bringing together parts of planet embryos that already had differentiated interiors, their own core, and even their own atmosphere. The process of accreting matter from space has continued, though far more slowly, right up to the present day.

The process of planet-building from swarms of planetesimals was modelled in 1986 by American geophysicist George Wetherill. The timescales he derived have been checked in the early 21st century using radiometric dating. This gives an age of 11 million years for Earth to achieve 63 per cent of its final mass. Mars, on the other hand, probably formed in less than a million years, reaching its final size through accretion. Some planetesimal cores might have formed in just 500,000 years.

Although scientists modelling this process for growing planets have created virtual aggregates of dust about 1 cm (0.39 in) across, they haven't been able to understand how a dusty blob grows from this size to 1 km (0.62 miles) across.

Safronov's hypothesis was not widely accepted in the West until the 1980s. A further contribution, the Kyoto model from Japan, also took a long time to be recognized. Developed by a team of astrophysicists at the University of Kyoto, it introduces the impact of gas on the whirling protoplanetary disk, which creates drag and

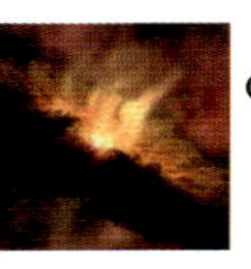

slows down the particles. This model allows scientists to describe the formation of the gas giants which Safranov's model struggles to explain.

An artist's impression of the early solar system, with the Sun forming as a star surrounded by a protoplanetary disk of planets, rocks, dust and gas.

GAS AND ROCKS

The composition of a planet depends on where it forms in the protoplanetary disk. Rocky planets form close to the star, as the materials they are made of condense to solids at relatively high temperatures. Large gas planets are made from more volatile materials that can only condense

COMETARY CLASHES

The French naturalist and mathematician Georges-Louis Leclerc, the Comte de Buffon, had a novel explanation for the formation of the solar system. He suggested in 1749 that comets crashed into the Sun, sending chunks spinning off into space; these then pursued independent lives as planets. In 1796, Laplace showed that this theory was not viable and any planets formed in this way would eventually fall back into the Sun. In fact, comets are far too small to have any effect on the Sun even if they all launched a dedicated onslaught together.

at much lower temperatures beyond what is known as the 'frost line' or 'snow line'. Earth's proximity to the Sun is explained by its composition of mainly non-volatile materials, largely silicate rock and iron.

Growing together

The formation of a star and its planets can happen simultaneously. A nascent star goes from being a proto-star to a type called T-Tauri star, that hasn't yet begun nuclear fusion, and finally to a fully-fledged star on the main sequence. The discovery in 2010 of a planet forming around a T-Tauri star showed planets can begin to grow while the star is still forming. Observations of disks of dust around other stars has greatly contributed to the understanding of how Earth and the other planets of the solar system probably formed.

This illustration of a young star reveals the complex structure of the protoplanetary disk, with concentric rings of gas.

The matter of Earth

The solar nebular disk contained everything from which Earth was made. The constituents of other bodies in the solar system which formed at the same time can tell us something about the raw materials of our own planet. For a long time, our only access to such material was meteorites – chunks of rock or metal that

fall from space. This has changed in the last 50 years; we can now collect matter directly from space, including from comets and asteroids.

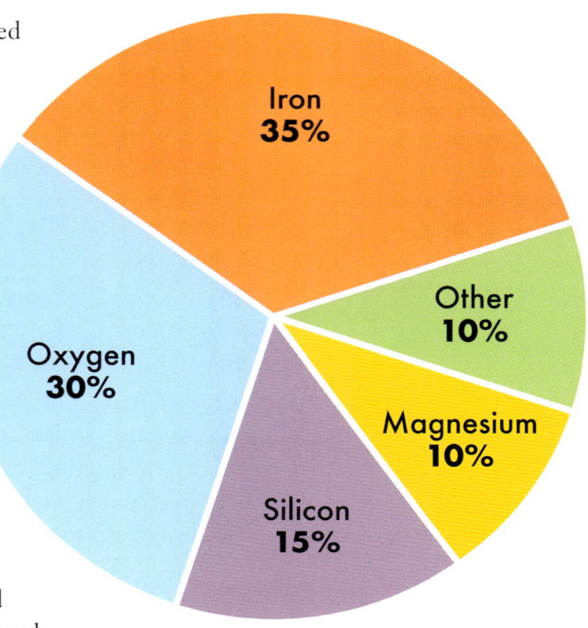

The approximate chemical composition of Earth, by mass.

Planetary leftovers

Meteorites can come from another large body (such as the Moon or Mars) or may be chunks that formed at the same time as the planets and have been circling the Sun unchanged for the last 4.6 billion years. Around 86 per cent of meteorites are of a type called chondrites. They condensed directly from the solar nebula and therefore represent the primordial material of the solar system.

Chondrites are all made of grains of rock and dust, but their composition differs and reflects where in the solar system they formed. Those forming furthest from the Sun are rich in carbon-containing substances (such as carbonate rocks), oxides and water. Those forming closest to the Sun, probably closer than the orbit of Mercury, have a high iron content.

Fingerprinting stardust

Chondrites consist of tiny spherical grains called chondrules embedded in dust, some as small as a few micrometres (millionths of a metre) across. A small proportion of chondrules are pre-solar grains – tiny flecks of matter that predate the Sun and come from interstellar space. Some pre-solar grains contain stardust – matter originally blasted out of stars or supernovae in the form of vapour, which then condensed in space. They include nano-sized diamonds, flecks of graphite, and silicates. Stardust grains can, at least in theory, be traced to a particular star.

Lying in layers

The meteorites that reveal the material from which Earth formed are the same all the way through. The same is not true of Earth, which is layered. At some point, the large collection of planetesimals changed into a single body and differentiated. The key to how it did it is heat.

The lightest part of Earth is its atmosphere, the cloak of gases above the

surface. The relatively thin layer we inhabit is a hard, rocky crust which carries the continental landmasses and the oceans. Beneath it is a thick layer of very hot rock, semi-liquid in places, called the mantle. The crust floats on top of this. The heaviest part is the core, which is made of iron. The iron core did not form first, but accumulated from material already in the planet after it had coalesced. Geoscientists have two possible explanations for how Earth's core became differentiated from the rock.

As the Earth grew larger, pressure at the centre increased, raising the temperature. Radioactive decay contributed to heating, and a cloaking atmosphere trapped the heat so the temperature continued to rise. When the temperature became high enough to melt the collection of silicate rock and metal particles which had collected, the heavier, metallic material gravitated inwards and the silicate rock surrounded it. As the planet cooled, it was left with a metallic core and a rocky exterior.

Many asteroids and comets, such as comet 67P/Churyumov–Gerasimenko shown here, are 'rubble piles', collections of chunks that have been brought together under gravity. Unlike Earth, they have not become hot enough to melt and reform as a single solid lump (a monolith).

Another theory suggests that when the temperature was a little lower, so the metal was still molten but the rock had solidified, liquid iron could percolate down through the rock and add to the core.

In 2013, Wendy Mao led an experiment at Stanford, USA, emulating conditions in early Earth. She subjected tiny portions of silicate rock and iron to the pressures and temperatures present in the early days of the planet: 64,000 times atmospheric pressure and 3,300 Kelvin, about 3,576 °C (6,469 °F). X-ray tomography of the particles showed that the molten iron formed an interconnected network. This could percolate down to the core from areas in early Earth where conditions were suitable.

It's possible that both mechanisms were at work, contributing to an iron-rich core when Earth was entirely molten and after it had begun to solidify.

Making a moon

Earth does not travel alone through space; it has a companion, our Moon. The Moon has always been visible to humans and its movements were tracked even in prehistoric times, as we can tell from the survival of artifacts and monuments that serve as lunar calendars. Although some cultures had myths about the creation of the Moon and the Earth, until the 19th century there was little or nothing in the way of scientific hypotheses about how it had formed.

Leaving aside the possibility that it was created by a god of some kind, there have been four theoretical models for how the Moon might have formed: a bit of Earth detached itself; the Earth and Moon formed at the same time; the Earth captured a pre-formed Moon; Earth suffered a collision with something and the Moon formed as a result.

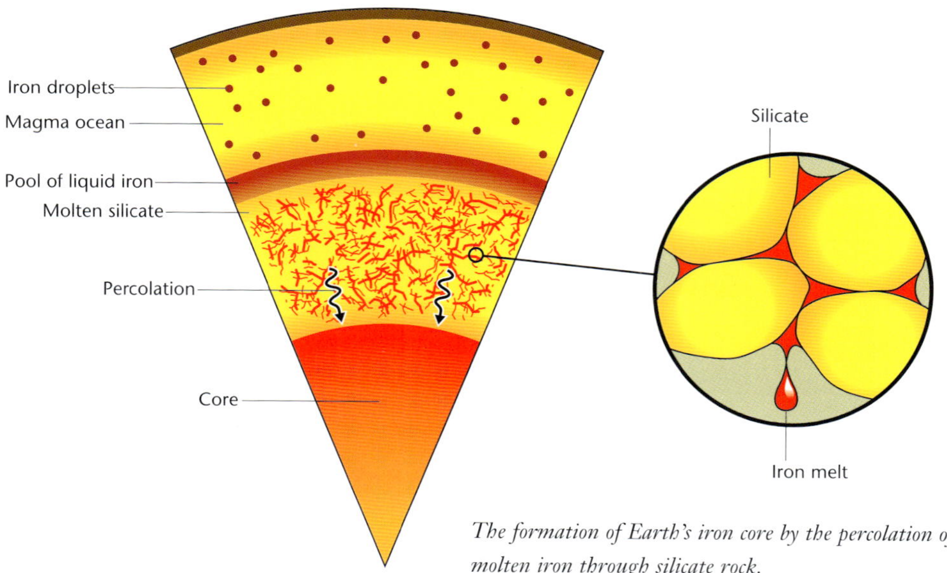

The formation of Earth's iron core by the percolation of molten iron through silicate rock.

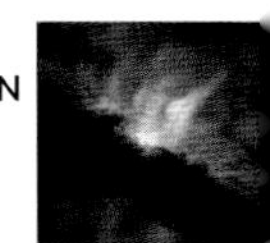

The first idea, that part of Earth broke away and formed the Moon, was proposed by George Darwin, son of the more famous Charles, in 1879. This is now called the fission theory. George Darwin suggested that Earth was spinning much more rapidly than previously thought, and that the gravitational pull of the Sun at the equator, coupled with centrifugal force, was sufficient to make Earth bulge outwards. A chunk of it broke away and went on to form the Moon. The Pacific Ocean seemed a likely scar to have been left by this injury to the planet (this idea was proposed by the English geologist Osmond Fisher in 1889). Fisher was a talented and prescient geologist, but he was wrong in this instance.

George Darwin's theory enjoyed support until the early 20th century when maths caught up with it. In 1909, Moulton showed that the current angular momentum would not be enough to favour a moon being spun off. In 1929, James Jeans demonstrated that early Earth would have needed to spin so quickly to cast off a moon that its days would have been only two hours and 39 minutes long. It would have been difficult to get from barely three hours to the current diurnal length of 24 hours, even over billions of years. In the 1960s, the early rotation period of Earth was fixed at 10–15 hours – much too slow for fission.

Until the mid-20th century, various geologists tweaked aspects of this model to make it more feasible. In a later version, the Moon as it escaped Earth had nine or ten times its current mass, but was so hot that much of it vaporized. Another explanation had it that there was a cloud of rocky debris around the Earth; some of this matter was recaptured by the planet and the rest escaped the system.

Bits and pieces

The second model describes the Moon and the Earth forming at the same time by the same method – accretion from the nebular disk. However, if the Moon and Earth accreted at the same distance from the Sun at the same time, we would expect them to be even more similar than they are.

The third model posited that the Moon had not directly formed in orbit around Earth but accreted elsewhere in the solar system. It had subsequently been captured by Earth's orbit. Many of the moons of the gas giants had been captured rather than formed *in situ* – but they are much smaller than Earth's moon.

Everything changed in the 1960s. The first photos of the far side of the Moon, returned by the Soviet spacecraft Luna 3 in 1959, showed that it was very different from the Earth-facing side. Then the Apollo Moon landings returned more data about its surface and even brought back samples of Moon rock and regolith (the dust that covers its surface). At last, scientists could work directly with Moon material to determine its precise composition.

Giant impact theory

New information brought a new theory – or, rather, resuscitated an old one. In 1946, Canadian geologist Reginald Daly had tried to fix problems with Darwin's spinning-off model by suggesting that an impact had knocked out a chunk of Earth which then became the Moon. Daly's idea was largely ignored until 1974 when

American astronomers William Hartmann and Donald Davies revived it as part of a more sophisticated scenario.

They proposed that a Mars-sized planet had formed nearby in the solar system on an orbit that crossed Earth's orbit. When the inevitable collision took place, the consequences were literally Earth-shattering. The smash released 100 million times as much energy as the asteroid strike that killed the non-avian dinosaurs 66 million years ago (see page 179). The impactor (now named Theia) vaporized, along with a substantial portion of Earth's mantle. The material mixed together, some falling back to Earth and recombining with the mantle and some solidifying in space and forming a debris ring from which the Moon accreted. This would explain why the Moon and Earth have a similar composition, and why the Moon has a very small core, as Earth's core would not have been vaporized in the collision. The name Theia is taken from Greek mythology: Theia was the mother of Selene, goddess of the Moon.

The giant impact hypothesis was not immediately popular, but after a conference in 1984 to compare the possible models, support for it rose to near consensus. Alastair Cameron was working on a giant impact model at the same time as Hartmann and Davies and developed one in which a tangential impact, with Theia striking Earth at an angle, produced the right conditions for the Moon to form.

However, in 2016, new work suggested a direct hit. Further refinement in 2019 by a team at the RIKEN Center for Computational Science in Japan showed that if Earth was still a sea of hot magma while Theia was a solid body, the Moon would have formed largely from material derived from Earth. If, on the other hand, Earth was already solid, it would have formed largely from Theia. As the Moon's composition is a close match for Earth and includes at least a small iron-nickel core, it's more likely that Earth had not solidified at the time of the impact. The impact is put

The giant impact hypothesis solves most of the main problems other models throw up, including explaining the evidence of catastrophic heating found in samples of Moon rock.

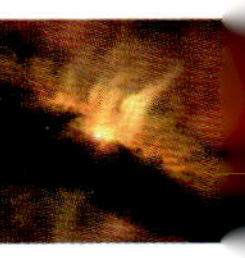

at 50 million years after the formation of the solar system, when Earth could plausibly still have been molten.

The new Japanese model has the Moon made of 80 per cent Earth-sourced matter and 20 per cent Theia-sourced matter, reversing the proportions given by earlier models. In the magma ball that would become the Moon, the heavier materials would sink inwards to form a small metallic core, as happened in early Earth.

Team Terra

With the giant impact hypothesis accepted by most planetary scientists, the situation about 4.5 billion years ago was that Earth had its natural satellite in place, had the mix of material it would keep (give or take a few meteorites, asteroids and comets adding to the mix), had begun to differentiate with a metallic core surrounded by a rocky mantle, and was ready to go forward to its life as a fully-fledged planet.

The newly formed Moon was much closer to Earth than now; it would have loomed large in the sky. It has slowly moved further away, and is still distancing itself at a rate of about 4 cm (1.6 in) per year. The rate varies as the arrangement of land and ocean on Earth changes.

CHAPTER 2

Long, long **AGO**

'The result, therefore, of this physical enquiry is that we find no vestige of a beginning, no prospect of an end.'

James Hutton, geologist, 1788

Just how old is Earth? It has taken a long time to answer this question with any degree of confidence. For many years, people attempted to calculate Earth's age based on dates in the Bible. Edmund Halley suggested that it could be encoded in the salinity of the oceans. Aristotle's view that Earth had always existed and was eternal later gained support among 19th-century geologists. But it was not until the astonishing discoveries of the 20th century that scientists began to realize the precise age of Earth and, consequentially, of the entire solar system.

The evidence of rocks and geological events (such as the meteor strike that created this crater) have revealed the age of Earth.

RELIGIOUS GUESSWORK

The Bible offers a mathematical method for calculating the date of Creation. The Book of Genesis gives the ages of a string of patriarchs, all of whom supposedly lived to nearly 1,000 years old; from these it is possible to work backwards from the birth of Christ to the supposed moment of Creation. Around 1650, James Ussher, Archbishop of Armagh and Primate of All Ireland, calculated that Creation had begun on 23 October, 4004 BC, at 6 pm. He claimed that time itself had begun the evening before, in a pre-Creation event.

Ussher was not the only Christian to attempt this calculation. The English Benedictine monk known as Venerable Bede settled on 3952 BC, the scientist Isaac Newton on 4000 BC and the astronomer Johannes Kepler calculated 3992 BC. It was not an exclusively Christian preoccupation: the 2nd-century Jewish rabbinic sage Jose ben Halafta set the date of Creation at 3761 BC.

Times past

Although the ages calculated from the Bible held sway in the West for a while, they could not contend with new directions in science which began within a decade of Ussher's dating. As we shall see, geologists began to realize that the landscape and continents were shaped by slow processes, an understanding which can't be reconciled with the belief that the world is only a few thousand years old.

Earlier thinkers, including the Ancient Greek philosopher Aristotle (383–323 BC) and the Renaissance polymath Leonardo da Vinci (1452–1519), suspected that Earth is quite old as they had worked out that fossils are the remains of ancient animals, often of types no longer living. But they did not attempt to calculate the age of the planet and would have had no way of doing so. Conversely, in the 1st century BC, the Roman poet Lucretius thought that Earth must have formed quite recently because there were no historical records from a time pre-dating the Trojan War.

Scientific approach

In the 1660s, the Danish scientist Niels Steensen, usually known as Nicolas Steno, laid the foundations of geology with his hypothesis that rocks are deposited in layers (called strata) with the oldest at the bottom (see page 81). Steno made no attempt to date the layers of rock, but cast into doubt the idea

Nicolas Steno was an anatomist as well as a geologist before becoming a bishop in later life.

that Earth was created complete and in its current form, and in a matter of days by a divinity with other work to do. Steno didn't intend to challenge the Bible; he was happy to assume that fossiliferous rocks were laid down during Noah's flood. He abandoned science in 1667, later becoming a bishop.

Salty seas and layers of rock

In 1715, the astronomer Edmund Halley attempted to use the salinity of the sea as a way of calculating how much time had passed since the formation of Earth. Halley noticed that rivers are fed by streams which sometimes spring from the ground and pour their water into the sea, carrying minerals dissolved from the rocks out to the oceans. He reckoned that if the sea started with zero salinity (not a valid assumption) and became salty over the period of its existence at a steady rate (another invalid assumption), it would be possible to work out Earth's age – but only if he knew the rate of accumulation of salt (which he did not).

The Russian polymath Mikhail Lomonosov (1711–65) was perhaps the first person to attempt scientific dating. In his *On the Strata of the Earth* (1763), Lomonosov made some astonishing discoveries and predictions – he detected the atmosphere of Venus and explained the formation of icebergs – but his work on the age of the planet was not his greatest achievement. He decided that Earth was created several hundred thousand years before the date accepted for the rest of the universe.

The French naturalist and mathematician Georges-Louis Leclerc, Comte de Buffon

The ice that forms an iceberg today often fell as snow tens of thousands of years ago.

(1707–88), tried to work out Earth's age by experimental means. He made a small globe similar to Earth in composition (as far as he knew it) and measured the rate at which it cooled. He believed that the planets had all formed from material knocked out of the Sun by a catastrophic comet strike. As a piece of the Sun would be searingly hot, Earth started out hot then slowly cooled. Eventually, molten rock solidified into a hard surface and water condensed and rained down, forming the oceans. From his measurements, Leclerc estimated Earth's age at around 70,000 years.

However, as Leclerc knew neither the starting temperature of Earth nor the rate at which it would cool in space (rather than at room temperature in France), his was a doomed attempt. But he had at least begun with the conviction that the Bible cannot be trusted to tell us about the geophysical history of Earth.

Georges-Louis Leclerc was the most important natural historian of the 18th century.

Many millions of years of rock strata are clearly visible in the Capitol Reef Monocline in Utah, USA.

While Leclerc's estimate of around 70,000 years was ten times the age the Bible seemed to indicate, by the mid-19th century geologists were arguing for a much older Earth. Looking at the rock strata and the fossils they contain, some scientists decided that Earth is indefinitely old. In 1838, the Scottish geologist Charles Lyell declared its age 'limitless'.

In 1876, geologist Thomas Mellard Reade returned to the idea of tracking dissolved minerals in the oceans, calculating that it would take 25 million years for calcium and magnesium sulphates to reach their current levels. He called the process 'chemical denudation', as water flowing through and over rocks was denuding them of mineral content and enriching the sea with it. Other people who repeated his calculations arrived at similar answers. In 1899, the Irish physicist and geologist John Joly came up with a precise age of 99.4 million years, though he later revised it to the wider range of 80–150 million years. In 1910, American geologist George Becker used the salt-clock method to calculate an age of 50–70 million years. The method doesn't work, of course. Not only does it assume a steady rate of accumulation starting from zero, but also that anything put into the sea stays there. Minerals in sea water are recycled through rock again, so the salinity does not steadily increase over time but stays much the same.

Another approach geologists used was to calculate from the rate at which sedimentary rocks are laid down. As rocks form in layers, knowing the rate of formation and the thickness of rock should make it possible to determine the rock's age. In the late 19th century, geologists first studied the rate of sedimentation and then applied it to the thickest rock deposits to calculate an age of 75–100 million years. Even if this method worked for a particular piece of rock, it is no use for calculating the age of an entire planet. Sedimentation doesn't happen at a steady, regular pace over millions or billions of years and all kinds of geological activity disrupt the layers and break up rocks.

Adding life to the mix

From the late 18th century, another group of scientists joined the debate about the age of Earth. Biology was beginning to accept that living things change and some become extinct over long periods of time. It became clear that imperceptible changes in organisms require a long time to build into significant developments. People were aware that organisms have not changed noticeably over hundreds or even a few thousand years, so some of the fossils of extinct organisms

being uncovered in the mid-19th century signalled very long timescales.

Physics gets in on the act

While geologists and biologists were steadily extending the period of time they felt Earth would have needed to reach its current state, another branch of science took a very different view. Physicists began to weigh in with ideas about materials and thermodynamics. It was a totally new approach.

In 1862, William Thomson (later Lord Kelvin) published his findings that Earth was between 20 million and 400 million years old. He based his figure on equations developed from the work of the French mathematician and physicist Joseph Fourier who, in the 1820s, laid the groundwork for the analysis of heat flows. Fourier believed that Earth had started out hot and was cooling. Thomson's equation calculated the age of Earth from three figures: the supposed original temperature of the molten rock of the planet; the geothermal gradient (the rate of increasing temperature with respect to increasing depth measured from the surface); and the rate at which heated silicate rock loses heat. At first, there were no figures for geothermal gradient, but by 1863 measurements had been made in several parts of the world. Using his equations, Thomson arrived at a second estimated age of 96 million years, but published his findings with a wider range to allow for uncertainty and variation in the thermal gradient and the thermal conductivity of Earth's rock.

Lord Kelvin (as he was by now) carried out a second calculation, this time to work out the likely lifespan of the Sun – since clearly Earth could not be older than the lifetime of the Sun. At the time, people assumed that the energy radiated by the Sun came from potential gravitational energy accumulated during its formation by accretion. In fact, as we have seen, the

Lord Kelvin was regarded as the greatest physicist of his day – he was not accustomed to being wrong.

energy of the Sun is provided by nuclear fission, but this process was undreamt of at the time. Kelvin calculated how much energy could have been stored up during accretion and worked out that it could sustain the Sun for no more than 100 million years. That tied in well with his figure of 96 million years, though it did leave the planet with a worryingly brief 4 million more years before its extinction.

Drawing 'reckless drafts on the bank of time'

The geologists did not like to see Earth's lifespan constrained by physics; Kelvin considered their approach to be highly unscientific and placed his faith in numbers and the laws of physics. This would have been fine if he had picked the right laws, but unfortunately they were not known and he was working from the wrong premise.

Nevertheless, Kelvin's calculations focused the minds of geologists and, in the words of American geologist Thomas Chamberlin, 'restrained the reckless drafts on the bank of time' which they had been drawing. After hearing Archibald Gerkie talk about the landscape of Scotland, Kelvin reported having the following conversation with British geologist Andrew Ramsay:

'I said . . . "You don't suppose geological history has run through 1,000,000,000 years?"
"Certainly I do."
"10,000,000,000 years?"
"Yes."
"The sun is a finite body. You can tell how many tons it is. Do you think it has been shining on for a million million years?"'

'The physicists have been insatiable and inexorable. As remorseless as Lear's daughters, they have cut down their grant of years by successive slices, until some of them have brought the number to something less than ten millions.'
Archibald Gerkie, geologist, 1895

Ramsay replied:

'I am as incapable of estimating and understanding the reasons which you physicists have for limiting geological time as you are incapable of understanding the geological reasons for our unlimited estimates.'

The geologists eventually settled for a very long, rather than infinite, period to have elapsed since Earth's formation.

Hot rocks

In 1895, John Perry, a former assistant of Kelvin's, pointed out that his dating rested on the thermal conductivity of rocks near the surface, but might not reflect the conductivity of rocks at greater depth. If rocks deep within Earth had higher conductivity than rocks at the surface, the interior would also be cooling and providing a large supply of energy. This would lead to a much longer continuation of heat flux at the surface. It meant the Earth could be far older than Kelvin's calculations suggested.

Perry recognized that the conductivity of rock increases a little at higher temperatures but, more importantly, the composition of Earth changes with

Quartz sandstone pillars in Zhangjiajie National Park, China. Crystalline silicate rocks such as these have relatively high thermal conductivity.

increasing pressure. Earth's interior conducts heat better than its surface. Perry calculated that if the interior had perfect heat conductivity, Earth might be two billion years old; with less-than-perfect conductivity, it could be much older. The geologists could therefore be right, without Kelvin's calculations being wrong.

Light and shadows

Luckily another method of dating was about to be discovered based on the astonishing discovery that the atoms in rocks decay over time in a rigorously predictable pattern and at an absolutely steady pace.

In 1896, the French chemist Henri Becquerel discovered radioactivity. While running tests on whether fluorescence, phosphorescence and X-rays are the same, or related, phenomena, he had exposed naturally phosphorescent crystals of

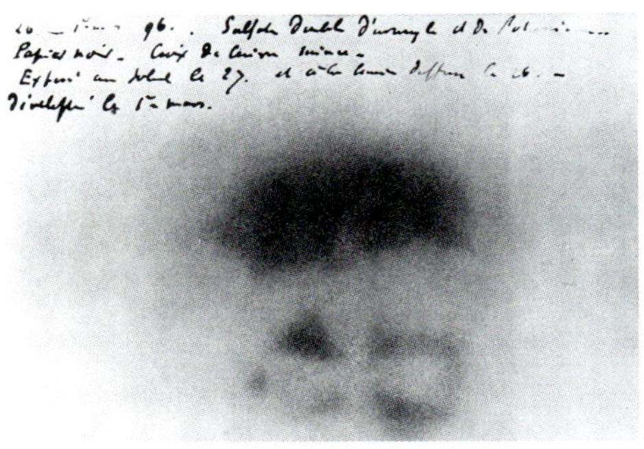

An abandoned experiment with phosphorescence led to Becquerel's accidental discovery of radioactivity.

potassium uranyl sulphate to sunlight, then put them on a photographic plate. He expected that they would absorb something from sunlight and re-emit it as X-rays, so marking the plate. He successfully developed some smudgy images of the crystals, so decided to investigate further. He planned a new experiment, but the weather was overcast, so he wrapped his crystals in a dark cloth and put them away in a drawer, along with his photographic plates and a metal cross. When he retrieved his equipment a few days later, he was astonished to find the photographic plate marked with the image of the cross, even though he had not exposed the crystals to sunlight. The Polish chemist Marie Curie coined the term 'radioactivity' for the phenomenon Becquerel had discovered.

In 1899, the New Zealand-born physicist Ernest Rutherford discovered that there are three different types of radioactivity, now called alpha, beta and gamma radiation. In 1903, Rutherford and the English chemist Frederick Soddy announced that radioactive elements break down predictably into other elements at a steady rate. This was mind-blowing: since the age of the alchemists, no one had supposed that an element could be broken down, made or changed.

The application of radiation to geology emerged early on. In 1905 Rutherford suggested that radioactive decay could be used to date rocks. In 1907, he defined the principle of 'half-life', which states that half a radioactive substance will decay in a specified period and this period is the same for all samples of the same radioactive element. The half-lifes of radioactive

HALF-LIFES – LONG AND SHORT	
Bismuth-209	19,000,000,000,000,000,000 years
Uranium-238	4.5 billion years
Lead-210	22.2 years
Radium-223	11.43 days
Uranium-240	14.1 hours
Francium-223	22 minutes
Carbon-15	2.45 seconds
Carbon-8	0.000000000000000000002 second

materials range from fractions of a second to billions of years (see box on previous page). The half-life of uranium is, coincidentally, about the same as the age of Earth, so only half of the uranium-238 that was present in the early Earth still exists today.

Dating Earth's crust

The American radiochemist Bertram Boltwood discovered that lead is always present in ores of uranium and thorium, and concluded that it must be produced by the radioactive decay of these elements. In 1907 he found there was more lead in the older uranium-bearing rocks, and worked out that he could use the uranium:lead ratio to calculate their age. He knew the rate at which uranium decays (its half-life), so could deduce from the proportions of lead and uranium present how long ago the uranium had been laid down. He had arrived at a new way of working out at least the minimum age of Earth's crust. The result of his calculation was 2.2 billion years. This was far older than Kelvin's figure and because it was based on analysis of the rocks themselves it seemed indisputable.

Decay chains

Three of the decay chains found in nature are useful to geologists because they enable us to measure the age of rocks. The chains are: the decay of uranium-238 (half-life 4.5 billion years) to lead-206 through 18 stages; the decay of uranium-235 (half-life 700 million years) to lead-207 through 15 stages; and the decay of thorium-232 (half-life 14 billion years) to lead-208 through ten stages.

Uranium-bearing rocks are of two main types: black shale and phosphorite. On the Cornish coast in the UK, fractured layers of mudstone and black shale are exposed at low tide.

Final countdown

Radiometric dating suddenly allowed Earth to be much older than geologists had dreamt possible. Arthur Holmes developed the uranium–lead method of radiometric dating and, in 1913, returned an estimate of 1.6 billion years for the oldest rocks (though not for Earth itself). He noted wryly that 'not many years ago, geologists were dissatisfied with the shortness of their time allowance; today they are confronted with an embarrassing superabundance'. In 1927 Holmes gave a new radiometric dating of rocks 3 billion years old; this was older than the universe was then supposed to be (about 1.8 billion years).

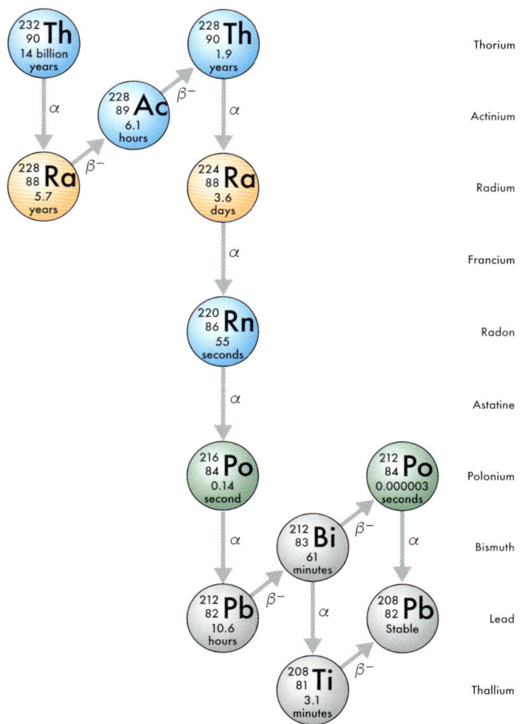

COLD EARLY EARTH?

The discovery of radioactivity in rocks allowed a possible explanation for Earth's heat other than that of Kelvin's hot-Earth-cooling model. An alternative model had rocks piling up in space and Earth accumulating as a cold lump, only to melt when there had been sufficient build-up of radiogenic heat after hundreds of millions or even billions of years. This could have produced the atmosphere rich in hydrogen, methane and ammonia suggested in the 1950s by Harold Urey (see page 120) and allowed the synthesis of complex organic molecules vital to life. The theory fell out of favour, however, after the heat involved in the collisions of accretion became irrefutable. Earth started hot; it didn't need to wait to warm up.

Thorium-232 has a very long half-life, but all other half-lifes in the chain are short, the longest being that of radium-228 (5.7 years). The ratio of thorium to lead is therefore a good enough indication of how much time has passed. If a sample is 87.5 per cent thorium and 12.5 per cent lead, around 3.5 billion years have passed since the rock formed: half would decay in 14 billion years, a quarter in 7 billion years, and an eighth (12.5 per cent) in 3.5 billion years.

Finally, in 1953, the American geologist Clair Cameron Patterson measured lead isotopes in the Canyon Diablo meteorite and returned the age of Earth as 4.53–4.58 billion years. Further radiometric dating of meteorites and of Moon rock retrieved by the Apollo landings has given an age of 4.54 billion years for Earth and 4.6 billion years for the solar system as a whole.

Slicing up time

The unwieldy timescale of 4.54 billion years makes it difficult to date events in geological time with any degree of certainty. Consequently, geologists developed a system of dating which names intervals in sequence. A relative sequence remains valid and useful even if the precise dates of events are unknown or change.

With the advent of radiometric dating methods it has become possible to fix approximate dates to the periods of Earth's history, and these have moved as methods have improved. The Cambrian Period, for example, is now considered to have started 541 million years ago (mya) but it was previously set at 542 mya (in 2009), 543 mya (in 1999) and 570 mya (in 1983).

Geological time is now divided into four eons, defined by the International Commission on Stratigraphy (ICS): Hadean, Archean, Proterozoic and Phanerozoic.

YOUNGER ↑ / OLDER ↓

Eon	Era	Period		
				◄ Today
Phanerozoic	Cenozoic	Quaternary		
		Neogene		
		Paleogene		
				◄ 66 Mya
	Mesozoic	Cretaceous		
		Jurassic		
		Triassic		
				◄ 252 Mya
	Paleozoic	Permian		
		Carboniferous	Pennsylvanian	
			Mississippian	
		Devonian		
		Silurian		
		Ordovician		
		Cambrian		
				◄ 541 Mya
Proterozoic	~	~		
				◄ 2.5 Bya
Archean	~	~		
				◄ 4.0 Bya
Hadean	~	~		
				◄ 4.54 Bya
Chaotian	~	~		

EONS OF TIME

A lack of physical evidence means that the Hadean eon has not traditionally been subdivided. However, in 2010 the American geologist Colin Goldblatt proposed that it be divided into three eras and six periods. He also suggested a new eon called the Chaotian be added before the Hadean. The Chaotian eon covers the time when Earth was forming in the protoplanetary disk. This new division separates events at the level of the solar system from events related only to the evolution of Earth.

Goldblatt proposed that the Hadean eon should start with the formation of the Moon and end at the same time as the hypothetical Late Heavy Bombardment. He also recommended naming the proto-Earth (Earth before the formation of the Moon) Tellus, after the Roman goddess of the Earth.

These are subdivided into eras, then periods, then epochs and, finally, ages.

Out of Hades

The first named geological eon is the Hadean, named after the Greek god of the underworld, Hades, in recognition of the hellish conditions that were presumed to have prevailed. It extends from the formation of Earth and the Moon until four billion years ago. In this period, Earth developed its solid, cool crust, formed oceans and an atmosphere, and made some of the earliest specks of rock that survive.

Following on from the Hadean, the Archean eon spans the period between 4 billion and 2.5 billion years ago, starting around the time Earth gained a stable, solid

Right: *A piece of gneiss (a metamorphic rock) from the Slave craton in Canada, some of the oldest exposed rock in the world. This dates from 4.03 billion years ago.*

Opposite: *A geological chart dating from Earth's beginnings and showing the division of time into eons, eras and periods.*

surface and ending around the time the atmosphere became oxygenated.

The oldest exposed rock formations date from the Archean eon, with just a few tiny grains considered to be Hadean. Earth's first continents are also thought to have formed in the Archean, centred on these islands of rock (see page 57). Life probably started during the Archean or even at the end of the Hadean (see page 122).

THE 'FAINT YOUNG SUN' PARADOX

During Earth's first billion years the Sun was about 15 per cent smaller than it is now and produced less heat. The result of this should have been that Earth was so cold that liquid oceans could not exist. Yet we know that Earth was not frozen but warmer than it is now. The flow of heat from Earth's interior to the surface, a mix of residual heat from accretion and heat from radioactive decay, was about three times the current rate. But interior heat provides only a tiny portion of Earth's heat; most of it comes from solar radiation.

The probable explanation is that Earth had a blanket of greenhouse gases, probably carbon dioxide and methane, which kept heat trapped near the surface. These gases would have been produced by volcanic activity or from asteroid impacts striking the Earth, or both. The greenhouse effect would have kept Earth warm enough to have liquid water and foster life.

LIVELY PLANET

The Proterozoic eon spans the period from 2.5 billion to 541 million years ago. During this time, Earth's continents formed, broke up and reformed several times as a result of tectonic activity (see pages 97–100). Life became prolific in the oceans, but was still limited to primitive algae and other microorganisms on land. It survived some extreme changes in climate and atmosphere – but only just. The Proterozoic eon began at the time when oxygen was first added to the atmosphere in large

Hallucigenia, *named for its bizarre appearance, was a velvet worm with spikes all along its back. Around 25 mm (1 in) long, it crawled across the seabed more than 500 million years ago.*

quantities. The Ediacaran Period at the end of the Proterozoic (635 to 541 mya) saw the evolution of multicellular soft-bodied organisms which left the first obvious fossils.

The Phanerozoic eon runs from 541 million years ago to the present. Not surprisingly, it is the eon for which we can calculate the most precise absolute dates. During this time, life spread over land, at times occupying all areas of the planet, and evolved into sophisticated advanced forms while maintaining a wide diversity of microbial citizens.

In the late 18th century, geologists began to allocate Earth's rocks to theoretical periods. They named these periods Primary, Secondary, Tertiary and Quaternary (the most recent). Their model was based on the principle of superposition – that the oldest rocks are the lowest stratum – explained by Steno (see page 81). The current period is still referred to as the Quaternary, though the other names have fallen out of use.

From the early 19th century, geologists such as William Smith in Britain and Georges Cuvier in France recognized that strata of rock could be identified and dated relative to each other by the fossils they contain (see Chapter 7). If a particular fossil is found in a particular stratum, wherever it is in the world, that fossil dates the stratum

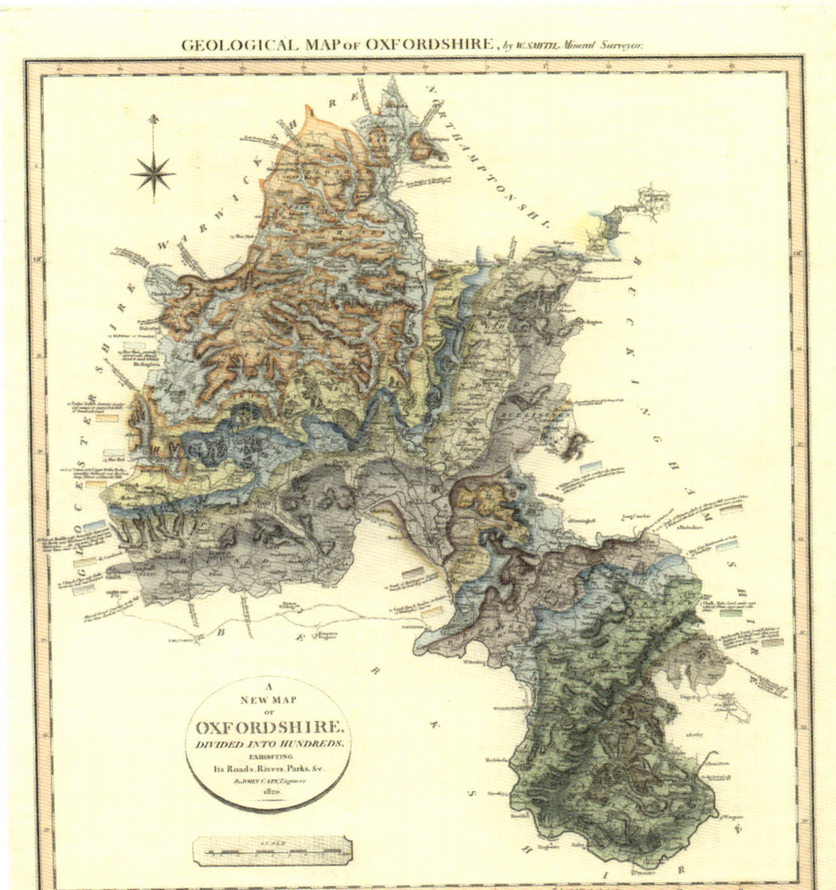

William Smith's geological map of Oxfordshire was part of his great undertaking to produce a colour-coded geological atlas of England, Wales and Scotland.

WHAT'S IN A NAME?

Systematic classification of geological time was begun by English geologists Adam Sedgwick and Roderick Murchison in the mid-19th century. Within each era, periods were named after the locations where distinctive rock strata had been found, or after the type of rock discovered. So 'Cambrian' is from the Roman name for Wales, *Cambria*, and 'Cretaceous' comes from *creta*, the Latin for chalk. 'Silurian' is from the name of an ancient Welsh tribe, the Silures.

The naming of the Cambrian and Silurian periods led to Sedgwick and Murchison falling out. They had been working together on the geology of Wales, and while Sedgwick defined the Cambrian Period, Murchison defined the Silurian Period (which comes later). Murchison used fossils extensively in defining his period, while Sedgwick did not. Working with different methods, there was some overlap in their definitions. Murchison claimed at first that part of the lower Silurian Period was actually part of the Cambrian, and later amended the claim to state that all of the Silurian Period was part of the Cambrian. It was a significant distinction, as Silurian fossils were the earliest known at the time, and both men hoped to claim the start of life on Earth (as they saw it) for their own named period. The issue was finally resolved in 1879 by Charles Lapworth, a colleague of Sedgwick, who proposed that the disputed lower Silurian/upper Cambrian periods be separately named the Ordovician.

and can be used to assess the relative ages of strata above and below it.

In 1841, British geologist John Phillips published the first geologic time scale, which ordered strata in rock according to the types of fossils found within. He introduced the term 'Mesozoic' (meaning 'middle life') for the era between Paleozoic ('ancient life') and Cenozoic ('recent life').

It's all relative

The principle of superposition (that new layers of rock cover older layers) demonstrates how one rock layer or fossil is older than another, but it doesn't give exact dating. When William Smith and his friends Joseph Townsend and Benjamin Richardson noticed a distinct change in rock layers between the fossilized plants in one strata and the fossilized seashells in the next, they could say that the plants came first but they could not date them absolutely. This change, now recognized as the division between the Carboniferous and Permian periods, can today be precisely dated to 298.9 million years ago. The Carboniferous Period, with its tree fossils, was the age when coal deposits were laid down, and the Permian saw the first large land animals.

The first dates allocated to different periods represented a lot of conjecture, but even today our best estimates might still be refined or overturned by future developments. The Jurassic Period was once thought to extend from 148 million to 108 million years ago, but has now been fixed at 200 million to 148 million years ago. At present, the end of the Permian Period is placed at 251.902 million years ago – that's given to an accuracy of 1,000 years. Some events can be more accurately dated than others. The end of the Permian Period was marked by volcanic eruptions and an extinction event that killed most life on Earth, but that would have taken years. The end of the Cretaceous Period was marked by another extinction, but this time caused by an asteroid strike. In theory, it could be pinpointed to a particular afternoon.

GOLDEN SPIKES

The International Commission on Stratigraphy (ICS) is the international body responsible for fixing the divisions in geological time. It identifies the exposed bits of rock that represent the lowest (therefore the oldest) boundary points between ages, called Global Boundary Stratotype Section and Points (GSSP). They are marked with a 'golden spike'. Other rocks around the world can be calibrated against these markers.

The golden spike at the GSSP of the Ladinian stage (upper Middle Triassic) in the Italian Alps. The boundary is at the base of the limestone bed which overlies the conspicuous groove.

CHAPTER 3

Earth, air and WATER

'Where once was solid land, Seas have I seene;
And solid land, where once deepe Seas have beene.'
George Sandys, *Ovid's Metamorphosis*,
Book XV, English translation, 1632

In its earliest days, Earth was a spinning mass of red-hot, semi-molten rock and metal. As it cooled, it hardened, eventually developing a solid outer crust with oceans of liquid water. In its first half billion years or so, it gained an atmosphere, oceans, a rocky surface – and probably life.

An artist's impression of an Archean landscape, with active volcanoes, stromatolites forming in shallow coastal waters and the Moon much closer to Earth than it is today.

As Earth cooled, much of it would have been covered with an ocean. The landmasses that formed bore no relation to the modern continents.

Earth's atmosphere

In its very early days, Earth was a place of searing heat and turmoil. These conditions have traditionally been depicted as continuing unabated for millions of years, the planet pummelled by asteroids and meteors that re-melted any solid surface almost as soon as it formed. The first of the officially recognized geological ages, the Hadean eon – named after Hades, the god of the underworld in Ancient Greek mythology – reflected this picture. But recently this notion of a turbulent half-billion years has been questioned. A new model is emerging, of an Earth solid, cool and possibly even amenable to life from a far earlier time than previously imagined.

Earth's atmosphere today is nothing at all like its first atmosphere. It has been through immense changes in the past, including being entirely replaced. In the 1940s, American geochemist Harrison Brown recognized that Earth has had two distinctly separate atmospheres. The first was captured directly from the solar nebula (a primary atmosphere); a later atmosphere formed from the material of the planet itself (a secondary atmosphere). Brown worked this out from something which wasn't there.

The missing neon

In 1924, the English chemist Frederick Aston noted that there is very little neon in Earth's atmosphere compared with the likely composition of the solar nebula. (Neon is an inert noble gas in the same family as helium.) The Sun contains about equal portions of neon and nitrogen, but Earth has around 86,000 times more

nitrogen than neon. Aston noted that all the noble gases are under-represented in Earth's atmosphere and proposed that their very inertness spelt their doom. He suggested that because the atoms of the noble gases were unable to cling to other atoms and in this way increase their mass, they were bounced back to whence they came:

'In the hurly-burly of colliding bodies ranging in mass from atoms upwards, the atoms of the inert gases, unconstrained by irrevocable chemical combination and free to collide and rebound indefinitely, would inevitably gravitate towards the larger masses and forsake the less. On this view the earth's share of inert gases has been lost to the sun, though whether they still remain there unchanged is outside the question.'

Harrison Brown picked up on this and started to investigate planetary atmospheres. He began by calculating the ratio of neon to silicon (common in Earth's rocks), so he could compare the ratios of these two elements on other planets without the sizes of planets skewing his measurements. He compared the amount of neon on Earth to the amount of the other noble gases – argon, krypton and xenon – and found that while Earth has about a millionth the abundance of most noble gases of the rest of the cosmos, it has only a billionth the amount of neon. Brown recognized that the significant feature of neon in this context is that, after helium, it is the lightest of the noble gases. He supposed something must have caused the escape of the neon into space. He concluded that the phenomenon which had taken neon would also have taken all the other lighter gases.

'It would appear that during the process of earth formation, the mechanism was such as to prohibit the retention of an appreciable fraction of any substance that existed at that time primarily in the gaseous state . . .

'It appears that the earth's atmosphere is almost entirely of secondary origin, and it was formed as the result of chemical processes that took place subsequent to the formation of the planet.'

Harrison Brown, 1949

A BAD PLAN

Not all of Harrison Brown's ideas were sound. In 1954 he suggested that world hunger could be solved by pumping lots of extra carbon dioxide into the atmosphere to stimulate the growth of crops. He thought that burning at least 500 billion tons of coal would double the carbon dioxide in the atmosphere. It soon became obvious that this was not a very good plan.

In 1949, Brown proposed that Earth's original atmosphere was similar in composition to the Sun (mostly hydrogen and helium). It had captured this from the solar nebular, but had lost it early in its history. Neon had been stripped from the planet along with hydrogen and helium. While neon was easily removed, the reactive gases and the heavier atoms of the other noble gases remained in larger numbers. The gas giants are much larger than Earth and have a much greater gravitational force, so could hold on to their light gases.

Grabbing an atmosphere

It turns out that Earth's first atmosphere might have been necessary for other aspects of the planet's development. The question of how Earth acquired this vital envelope of gases was addressed in 1979, by Chushiro Hayashi of the University of Kyoto.

Any object of significant mass will draw material towards itself. Hayashi demonstrated that once Earth had acquired one tenth of its current mass, it attracted a considerable cloak of gas from the surrounding solar nebula. This would have been mostly hydrogen, but with some helium and other gases in smaller quantities. One result of this atmosphere was that as Earth's mass increased its surface temperature rose because of the blanketing effect of the atmosphere. By the time Earth had a quarter its current mass, the surface temperature would have been c. 1,500 K, hot enough to melt all its constituents. This allowed the heavier metals to sink towards the centre of the planet. Under its fully formed primary atmosphere, the temperature was probably around 3,000 K.

> *'During stages where the solar nebula was existing, the proto-Earth was melted almost completely and the melted metal sedimented towards the center to form the core of the Earth.'*
>
> Chushiro Hayashi, 1979

Stripped

This stage didn't last long, though. When the Sun began nuclear fission, solar radiation

Meteors that impacted early Earth brought with them extra material which contributed to the final composition of our planet.

stripped away the outer layers of the Sun and remnants of the solar nebula, carrying it outwards through the solar system. Earth's atmosphere was in a vulnerable position. Hydrogen and helium are so light that they are easily lost to space, and the solar radiation and wind probably carried away Earth's share of the gases, too. In Hayashi's model, Earth lost its first atmosphere from the top down over a period of around 100 million years.

But there are other possibilities: a major impact may have raised the temperature sufficiently to accelerate the hydrogen and helium to escape velocity (see box) and the atmosphere could have been lost entirely in a matter of hours, as suggested in 2006 by Kevin Zahnle of NASA. However, scientists at the Massachusetts Institute of Technology (MIT), Hebrew University and Caltech argued in 2014 that a single, massive impact would have generated enough heat to melt

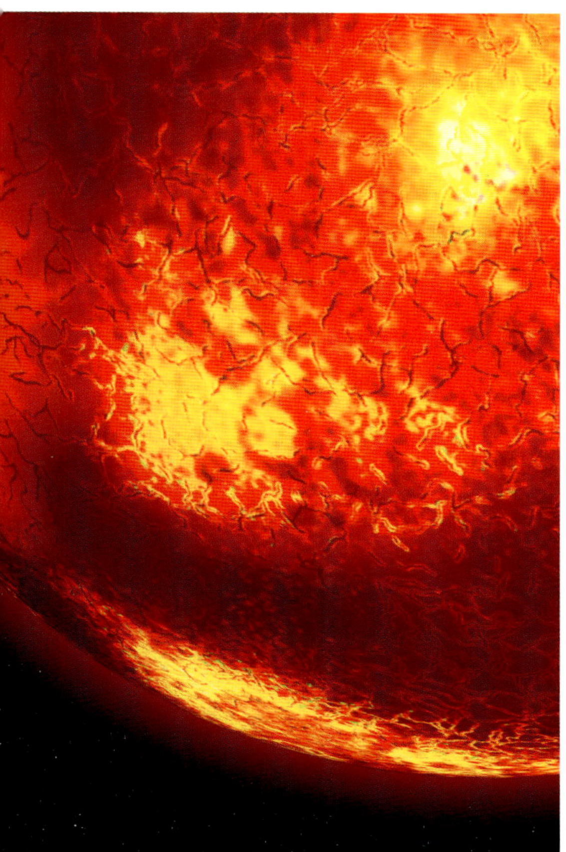

HOT, FAST AND FREE

The atoms or molecules of a gas move more quickly at a high temperature than at a low one; this movement is what defines heat. A smaller atom, supplied with energy, will move faster than a large atom supplied with the same amount of energy.

To escape from the gravitational pull of an object, a gas molecule needs to achieve escape velocity. On Earth, escape velocity is 11.3 km (7 miles) per second. If a gas is hot enough and light enough that its molecules move at more than 11.3 km per second, they can escape from Earth. Similarly, if gas molecules passing Earth are travelling more slowly than this, they can be captured and dragged into Earth's atmosphere.

At temperatures below 2,000 K, gases with molecular mass less than 10 will be able to escape, but those with molecular mass higher than 10 will be captive.

the interior of Earth. The current internal structure of Earth suggests this did not happen. Instead, these scientists propose that thousands of small impacts from asteroids and meteors punched through Earth's early atmosphere at around the same time as the formation of the Moon. These asteroids vaporized on impact, forcefully ejecting volatile substances which pushed out and replaced chunks of the atmosphere above the impact sites.

A home-made atmosphere

All was not lost with the first atmosphere, though, as Earth accumulated a new, secondary atmosphere. Our current atmosphere developed from this primordial version. Its composition has varied considerably over more than four billion years, evolving with geological and, later, biological changes.

The ingredients for the new atmosphere were already waiting within Earth. As chondrites (non-metallic meteorites) and larger lumps swirled around in the solar nebula, volatiles clung to their surfaces. Hydrogen readily forms volatile compounds such as ammonia (with nitrogen), methane (with carbon) and water (with oxygen). As Earth formed from a collection of lumps and chunks, it accumulated volatile materials which clung to the outside of these building blocks. During the early phase of accretion, with low energy impacts between small particles and with relatively cool conditions, incoming materials held on to their volatiles, so they were locked into the forming planetesimal. Later, when embryonic Earth was larger, the higher temperature and greater energy of some impacts led to the volatiles being released immediately, making a proto-atmosphere. Impacts from comets, which are mostly ice, would always release volatiles immediately, because the ice would immediately melt and be vaporized in Earth's higher temperature.

An atmosphere from inside

As we have seen, the young planet was hot, heated by the gravity of the accreting

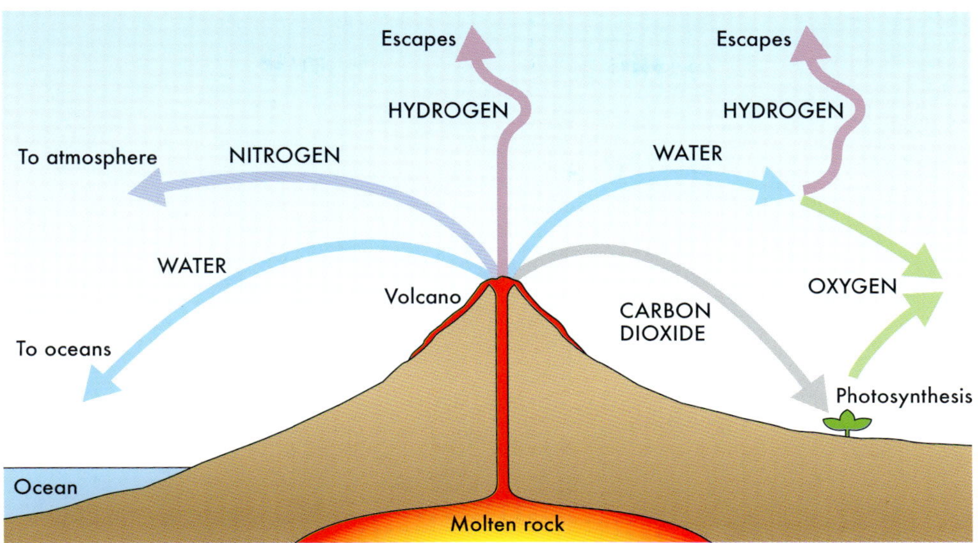

Facing page: *Gases still escape from the magma beneath the surface, bubbling up in volcanic fields such as this geothermal pool in Iceland.*

Above: *In Earth's early atmosphere, gases escape from the planet's molten interior through volcanic activity. Hydrogen escapes into space, but carbon, nitrogen and oxygen do not.*

parts, radioactivity, impacts, its cloaking atmosphere and radiation from the Sun. When the temperature was high enough to melt the rock, materials moved through it according to their mass; the heaviest metals migrated towards the centre and the lightest gases moved towards the surface. Volatiles locked within the forming planet escaped through the hot, semi-molten rock in a process known as outgassing. They rose to the surface through the magma, or through gaps in the hardened rock, and escaped to build a second atmosphere. As well as the volatiles, gaseous elements were present in compounds such as nitrates, oxides, sulphides and so on. Chemical reactions within the planet caused some of these to be released and to make their way to the surface. Most of the atmosphere generated from within was water vapour and carbon dioxide (see diagram above).

Always active

Earth's atmosphere is constantly changing, as elements are recycled in different forms. Hydrogen can be present in ammonia (NH_3), water (H_2O), methane (CH_4) and as molecular hydrogen (H_2), for example. Although molecular hydrogen is light enough to escape easily into space, methane, water and ammonia have greater molecular mass so do not escape at low temperatures. Instead they break down: intense ultraviolet radiation, such as that coming from the Sun, dismantles methane into carbon and hydrogen, water into oxygen and hydrogen,

and ammonia into nitrogen and hydrogen in a process called photodissociation, or photolysis.

Gaseous elements, like other elements, are constantly recycled. When an element becomes locked away in a 'sink' – for example, when carbon is locked into carbonate rocks – it is removed from the system for a time and the balance of the atmosphere changes. When a new source such as a massive volcanic eruption pours carbon dioxide into the atmosphere, the balance changes again. The cycles we have now for carbon, oxygen and nitrogen are not the same as those in place billions of years ago, as living things now play an important role in chemical cycles.

The surface of Venus is not visible, masked by a thick atmosphere that traps heat close to the surface.

Planetary comparisons

Earth's primordial atmosphere was rich in carbon dioxide. As we have found to our cost, carbon dioxide is a potent greenhouse gas. Four billion years ago, it trapped heat close to the surface, keeping Earth warm; without it, Earth would have frozen at this point. To see the difference such an atmosphere makes, we need only compare two of our neighbours in space, Mercury and Venus.

The closest planet to the Sun, Mercury is hot on one side and cold on the other. It has virtually no atmosphere to shield it from heat and radiation or to trap heat near the surface. It turns slowly, so each solar day on Mercury is 176 Earth days long – half a day spent facing a scorching star can warm the surface up to 430 °C (800 °F). Half a day spent facing away from the Sun can cool it to a minimum of -180 °C (-290 °F) at night. Although Venus is further from the Sun than Mercury, it is hot all the time. Reaching 467 °C (872 °F), the surface is hot enough to melt lead. There is little difference between temperatures day and night, even though a Venusian day is 243 Earth days long. The reason for this is Venus' thick atmosphere, made mostly of carbon dioxide, which traps heat near the surface and ratchets up the temperature over time. The same happened on the early Earth, though to a lesser degree. Earth had something Venus didn't have, and it changed the fate of the planet. Earth had an active surface, and water.

Many scientists believe that four billion years ago the early atmosphere of Earth was very like the current atmosphere of Venus: a

heavy, high-pressure atmosphere of carbon dioxide, possibly with sulphuric acid clouds, resulting in a surface temperature of 230 °C. Venus possibly had liquid water at that time, too, but it had no tectonic activity so lost its water and heated up.

Locking away carbon

When Earth's atmosphere was rich in carbon dioxide, this readily dissolved in the ocean and formed carbonate and bicarbonate ions. As the ions combine with calcium from chemically weathered rock washed in from rivers, they form carbonate rock on the seabed. The calcium comes from the weathering of surface rocks by rain. This is part of the slow carbon cycle. The carbon dioxide retrieved from carbonate rock is integrated into Earth's mantle and

SAVED BY THE SEA

Eventually, the extra carbon dioxide that humans have poured into the atmosphere since the Industrial Revolution will dissolve in the oceans and be locked up in carbonate rock. When this reaches the subduction zones at the edges of the oceans, it will be drawn back into the mantle. It will take thousands of years, as it relies on the slow process of exchanging water at the top of the ocean with water from the bottom. The 'ocean conveyor belt' which moves water around Earth and between the depths and the surface takes around 1,000 years for a full cycle, so this is not a quick fix for climate change.

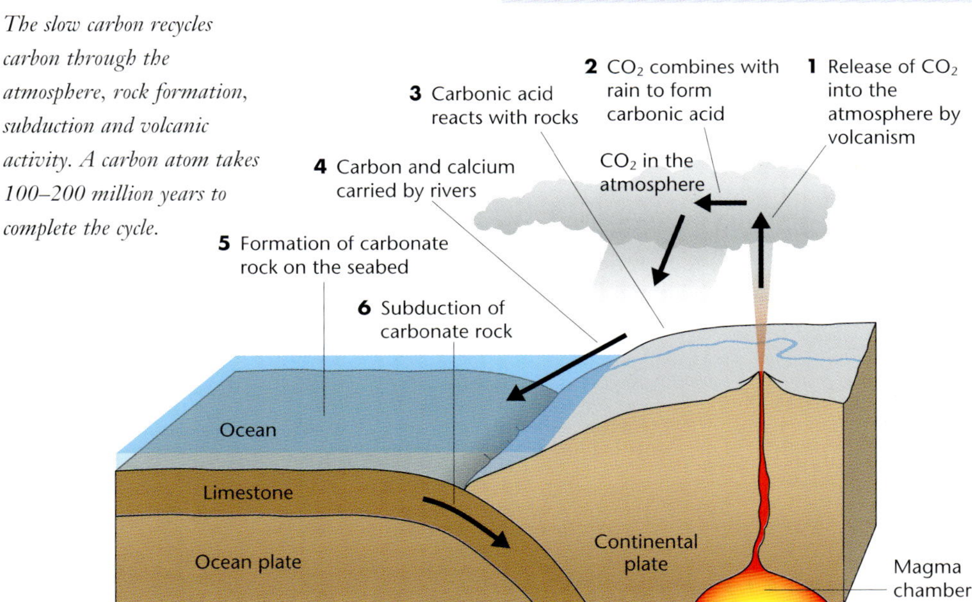

The slow carbon recycles carbon through the atmosphere, rock formation, subduction and volcanic activity. A carbon atom takes 100–200 million years to complete the cycle.

eventually returns to the atmosphere in volcanic gases.

As rock is subducted (drawn down into the mantle), new carbonate rock forms in its place. Deeply subducted carbonate rock would carry carbon dioxide into the mantle, locking it away for a long time. Today, carbonate rock forms in the top few hundred metres of young oceanic crust, but the main means of removing carbon dioxide from the atmosphere is by the photosynthesis of plants and algae. Before they evolved, carbon dioxide was only recycled geologically. A substantial amount of early Earth's carbon dioxide might have been removed in 100 million years by this process.

A Roman floor mosaic of the Greek god of the sea, Oceanus (right) with his consort the goddess Tethys. Oceanus was the originator and ruler of the seas. Tethys was the mother of rain clouds and their children were the gods and nymphs of rivers and streams, suggesting some understanding of the water cycle more than 2,500 years ago.

From rock to ocean

Like the origins of Earth itself, the question of where the oceans came from has been the subject of myth and legend as well as philosophical conjecture and scientific investigation. Three possibilities have been suggested as the source of water. Scientists suggest it may have come from within the rocks that formed proto-Earth; or from asteroids and meteors crashing into Earth; or from comets crashing into Earth.

Whatever the exact source, much of Earth's water is older than the Sun. It formed in interstellar space, where it floated around as ice crystals before being captured in the solar nebula.

To investigate which of the three sources is most likely (or whether all could have contributed), scientists have examined the precise chemical composition of the water found on Earth. Its origins can be indicated by the proportion of 'heavy water' – water made with 'heavy' hydrogen or deuterium (see box opposite) – that is present.

The proportion of heavy water found on Earth today is much lower than the proportion found in comets. The comets Halley and Hyakutake, for example, contain twice the amount of heavy water found in Earth's oceans. This means that all of Earth's water cannot have resulted from random comet strikes.

HEAVY WATER

All water molecules are made of two hydrogen atoms and one oxygen atom, giving its formula H_2O. But not all hydrogen atoms are identical. Normal hydrogen has a single proton in its nucleus, giving it an atomic mass of 1. A variant of hydrogen called deuterium also has a neutron in the nucleus, giving it an atomic mass of 2. In a water molecule, either or both hydrogen atoms can be replaced by deuterium. If only one is replaced, the compound is called semi-heavy water (and can be written HDO). Heavy water is D_2O. Heavy water has a slightly higher freezing and boiling point than regular water and is 10 per cent denser; heavy water ice sinks in regular water.

The possibility that it was delivered by meteorites colliding with Earth has also been ruled out, as the water in most meteorites contains quantities of the rare gas xenon – around ten times more than can be found in Earth's water.

In 2014 it was discovered that the ratio of heavy water on our planet matches that found in meteorites which have broken away from the asteroid Vesta and fallen to Earth. Vesta's composition was effectively 'locked' 14 million years after the start of the solar system when the asteroid froze, so it represents unaltered rocky material from the time when Earth was a quarter to half of its current size. As Earth and Vesta formed in the same region of the solar system, this discovery suggests that a substantial portion of Earth's water was trapped inside the planet as it formed.

There is still water trapped in the molten rock of Earth's mantle. In 1995, work in Switzerland by Peter Ulmer and Volkmar Trommsdorff revealed that minerals 150–200 km (93–124 miles) below the surface could contain water; recent computer modelling suggests water might even exist 660 km (373 miles) beneath Earth's crust.

The primordial atmosphere contained a large amount of water vapour. This condensed to form clouds, and when the conditions were right the clouds produced rain. Where the rain fell on to rocks where the temperature was less than the boiling point of water, it drained down to the lowest lying areas, where it pooled. Over tens of millions of years, these pools grew into oceans. Evidence from zircon crystals, the oldest surviving fragments of rock, indicate that Earth's oceans existed 4.3 billion years ago.

Rain probably fell for centuries to form the oceans.

Earth's early oceans would have looked exactly the same as they do today – this photo could have been taken 3.8 billion years ago.

The addition of chemistry

Chemicals dissolved into these large bodies of water from the rocks and the air, making the oceans acidic (from chlorine and carbon dioxide) and salty (from minerals). But, despite some variation over time, the volume and salinity of Earth's oceans has remained much the same. Salinity has risen and fallen but the sea does not consistently become saltier, as Edmund Halley supposed it did. The temperature of the oceans changes with the global climate, and they have both been hotter and colder than they are now.

A burning question

Earth cooled from the outside, but it was still hot beneath the surface. The first rock was probably basalt mush, perhaps 100 km (62 miles) thick, lying on top of a molten and semi-molten rock mixture known as magma.

The first rocks to solidify from the magma at Earth's surface were probably rich in magnesium and iron (ultramafic). They formed in patches of thin crust; as more magma was rising all the time, it would break the crust. This resulted in subduction zones, where lumps of crust were pulled down (because they were denser than the rising molten rock) and would then melt again. Mafic rock is high in iron; as this rock re-melted, some of the iron would sink. Over time, the crust grew to contain a lot of silica, forming lighter felsic rock.

There are other possible explanations for the formation of patches of enduring rock. One is that rock thickened from below, either by rising magma piling up and hardening beneath it or by subducted rock not melting but lodging beneath it (underplating).

Subduction along fractures in the emerging crust eventually produced the first island arcs – chains of volcanoes where the magma rises to the surface and hardens into solid rock. As this new magma was felsic, it did not easily subduct because it wasn't significantly heavier and denser than the underlying magma. Slowly, the island arcs

merged as they grew, accreting more rock at their edges until they became large masses of silicate rock. These are still visible as continental shield areas, large stable regions of low relief. Later, a 'platform' of rock was laid down around the shield, and together they formed a 'craton'. This development marked the start of the Archean eon.

Cratons help us to work out how continents have formed and reformed. These stable, central, primordial chunks (also described as 'shields') do not subduct and disappear beneath the oceanic crust, but they do weather and erode. There are around 30–40 cratons of varying sizes

A pre-Cambrian basalt lava ridge, part of the continental shield in Ontario, Canada.

EARTH'S OLDEST ROCK

While all the earliest crust has gone from Earth, some of it might have been found – on the Moon. In 2019, scientists at NASA discovered a splinter of rock they believe formed on Earth and was blasted out by a meteor strike, eventually landing on the Moon. The piece was brought back by the crew of Apollo 14 – it's made of quartz, feldspar and zircon, minerals which are common on Earth but very rare on the Moon. It was formed in conditions that have never existed on the Moon, but did on Earth four billion years ago. At this time, the Moon was a third of its current distance from Earth – a short hop away for impact debris.

The felsite clast (indicated) in this piece of moon rock is likely to be from a chunk of Earth rock that landed on the Moon as a meteorite.

around the world, but less than 10 per cent of current cratons are likely to have formed in the Archean. During the early Archean, there was a huge amount of tectonic and volcanic activity.

The word 'craton' comes from *kratogen* (from the Greek *kratos*, meaning 'strength') and was proposed by the Austrian geologist Leopold Kober in the 1920s; it was shortened to *kraton* by Hans Stille. Kober

SETTLED STATUS

Today, Earth has a solid, cold crust which is up to 50 km (30 miles) thick on the continental landmasses, but thinner and denser beneath the oceans. The mantle is made up of magma which is about the consistency of road tar and flows slowly. It is mobile enough for convection currents to move through it.

The upper mantle occupies about a quarter of the mantle's depth. Beneath it, the outer core is made of liquid iron, sulfur and some nickel; its temperature is 4,000–5,000 °C (7,200–9,000 °F). Right at the centre of the Earth is the inner core, made of the same metal mix, but solid. At 5,000–7,000 °C (9,000–13,000 °F), it is hotter than the outer core, but it is under such immense pressure that the atoms don't have space to move.

INSIDE EARTH

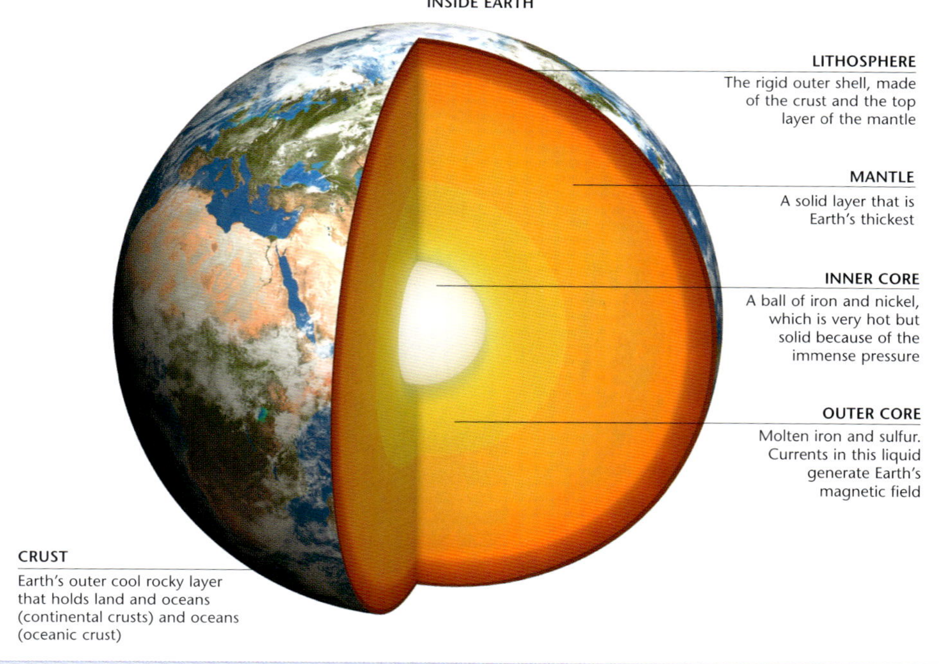

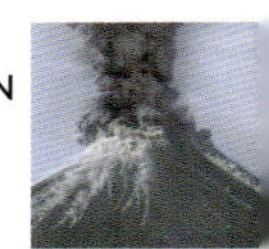

The distribution of pre-Cambrian cratons (dark orange) in the modern continents.

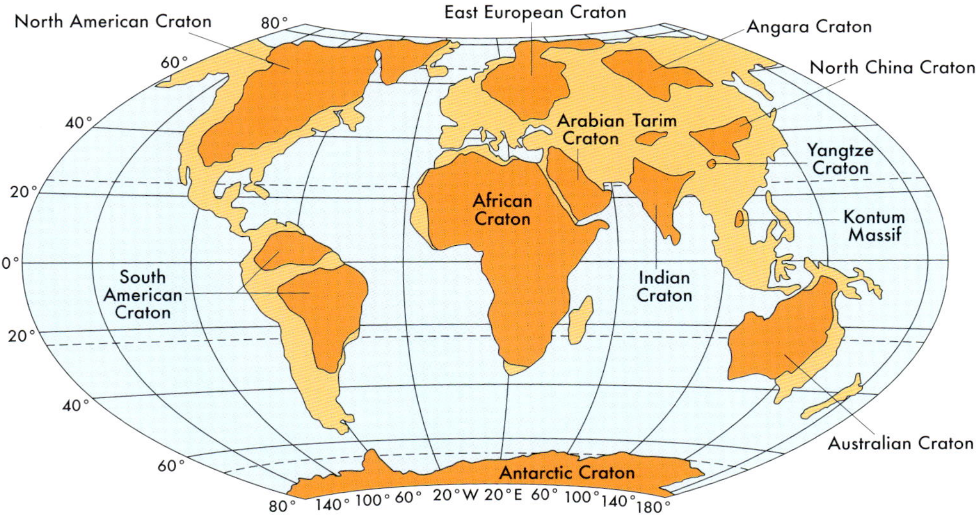

used it to describe the stable centres of continents, around which subduction zones form. He used the word *oregen* to describe an area where the surface was active and changeable. Kober and Still believed that the surface features of Earth – particularly mountains – were the result of the interior of the planet shrinking as it cooled. This caused the crust to wrinkle up to fit, since there was too much surface for a smaller planet to be enveloped evenly. This is part of a well-developed contractionist view of Earth's geology, developed by the English geologist Oswald Fisher in 1841. The theory was highly influential, but Fisher himself abandoned it in 1873, deciding that it couldn't account for the irregularities found on Earth's surface.

Making continents

There were no continents in the Archean, just cratons dotted around a global ocean. Later, as these cratons collided and combined, continent-building began. The continents continued to grow as more magma was deposited around their edges.

The first continent thought to have existed was named Vaalbara (in 1996) after the two cratons that formed it: Kaapvaal, now in South Africa, and Pilbara, now in Australia. They may have been joined together around 3.8 billion years ago, making a small continent which is nevertheless referred to as a supercontinent because it was the only significant landmass at the time (supercontinents need to contain at least 75 per cent of the world's landmass).

NAME	FORMED	BROKE UP	CURRENT LOCATION OF CRATONS
?Vaalbara	3.6 bya	2.8 bya	Southern Africa, northwest Australia
?Ur	3 bya	200 mya	India, Madagascar, Australia
Kenorland	2.7 bya	2 bya	North America, Greenland, Scandinavia, western Australia, the Kalahari Desert
Columbia/ Nena	1.8 bya	1.3 bya	Everywhere
Rodinia	1 bya	750-650 mya	Everywhere
Pangea	450-320 mya	185 mya	Everywhere

The existence of Vaalbara is speculative; some geologists favour Ur, a supercontinent thought to have formed about three billion years ago. Described as a supercontinent because it comprised most of the available land, Ur was only about the size of Australia.

After Ur, supercontinents came and went, their location and size now uncertain. The first uncontested supercontinent, Rodinia, formed around a billion years ago. It is thought to have consisted of Ur, with the addition of two non-supercontinents named Atlantica and Nena. There was possibly also a short-lived (60 million years) supercontinent named Pannotia between Rodinia and Pangea. It is thought to have formed when two large landmasses drifted alongside each other, instead of colliding with slow but relentless force in the usual manner of supercontinent-building.

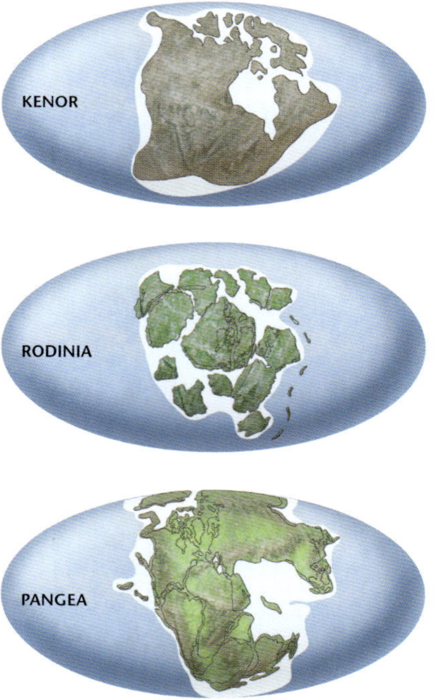

The supercontinents Kenor, Rodinia and Pangea formed as the land repeatedly broke up and melded together again.

Spotting supercontinents

The history of supercontinents only began to be pieced together at the start of the 20th

century when geologists first discovered continental drift. The existence of Pangea was proposed by Alfred Wegener in 1912; evidence for Rodinia began to emerge in the 1970s (the supercontinent was properly described in 1990); and Columbia was described in 2002.

The only landmasses we have today are non-supercontinents. If modern humanity was not influenced by geopolitics, we might consider North and South America as a single continent, with the massive block of Asia, Europe and Africa making another. It would take only the closing of the Bering Straits to create a supercontinent.

The most famous and recent supercontinents are Gondwana (now Africa, India, Madagascar, Australia and Antarctica) and Laurasia (Europe, Asia and North America), which resulted when the great supercontinent Pangea broke up. Gondwana pre-dates Pangea and was one of its forming landmasses. Gondwana finally broke apart between 140 million and 45 million years ago.

Not done yet

The formation and destruction of supercontinents continues today. The landmasses will continue to move for billions of years. Geologists predict that the next supercontinent might be formed by the closing of the Pacific Ocean, creating 'Novopangaea', while the closing of the Atlantic Ocean instead could create 'Pangaea Ultima'. A slightly different model has everything drifting northwards, the Americas growing closer together and colliding with Eurasia around the North Pole to form 'Amasia'.

An artist's impression of the coast of Kenor, with simple photosynthesizing algae in the shallow sea.

All together

The location and grouping of landmasses has a significant impact on Earth's climate and oceans. Obviously, the size and position of the oceans is dictated by the size and position of the land. The depth of the oceans is affected by temperature; and the composition of the atmosphere affects land and ocean. The oceans are also affected by tides created by the Moon, which has moved progressively further away from Earth over the last 4.5 billion years.

When all the land is collected together near the equator, as it was in the supercontinent Rodinia, Earth tends to be cool. This is because continental land reflects more heat back into space than ocean water does. Tropical rains, falling relentlessly on the land, eroded rocks and brought about chemical reactions which removed carbon dioxide from the atmosphere (see page 173). This added to the cooling effect until Earth was eventually plunged into the frozen state of a 'Snowball Earth' (see pages 125–6). The interaction between climate and landmasses had a significant impact on life, too, as we shall see in Chapter 6; in its turn, life has affected climate and landscape.

Deep within

The force driving Earth's evolution and the movement of the landmasses was heat from within. The first person to suspect this was James Hutton, in the 18th century. After years spent observing the structure of the land and the effects of wind and weather erosion, he concluded that the planet is hot on the inside, and that the heat produces forces which shift and deform the land. Hutton's insight that Earth might be hot under the surface was the first scientific approach to the possible internal structure of the planet.

Below: *Italian artist Sandro Botticelli (1445–1510) depicted a tiered Hell reaching deep underground, as described by the poet Dante.*

***Right:** Kircher's model of Earth had space inside for chambers and canals of fire and water.*

ON SOLID GROUND?

While it might seem obvious to us today that the Earth is solid, there are plenty of myths and religious stories that present a very different view, with notions of an underworld, or Hell. Even if Earth doesn't house an underworld, it's not clear from our observations of the surface whether it is the same all the way through.

The first proposal that Earth is not solid and homogenous came in 1664 with the work of German Jesuit scholar polymath Athanasius Kircher. His *Mundus Subterraneus* described a vast central fire at Earth's core. It supported the notion that the land beneath our feet lies on top of underground lava lakes and chambers (see pages 106–7). Although this seems prescient, he didn't get it all right: he supposed that water is sucked into a hole at the North Pole, heated at the central furnace, and expelled with force at the South Pole.

Edmund Halley holding a piece of paper depicting his model of Earth's interior.

In 1692, the astronomer Edmund Halley outlined a theory of a hollow Earth with an interior as an ornate arrangement of concentric shells with gaps in between, which he said explained the inconsistency of Earth's magnetic field. He reasoned there could not be anything moving around in the solid rock that would make the magnetic field uneven, so explained the phenomenon by dispensing with the rock. He theorized

GOD'S WATER TANK

In his *Sacred Theory of the Earth*, published in 1680–89, the Reverend Thomas Burnet maintained that Earth was the ruined remnant of a paradisiacal antediluvian world. He attempted to explain its history as recorded by scripture, but in scientific terms. He began with the premise that the biblical account of Creation was true, but the science suggested that there was not enough water on Earth to drown the planet in a global flood. He therefore decided there must be extra water somewhere, and the logical place was underground. He concluded that God had secreted a store of water beneath Earth's crust, in case there was a need to unleash a worldwide flood. The water would be released at the appropriate moment by creating a crack in the surface.

The frontispiece of Burnet's Sacred History *shows the history of the world, from chaos in darkness, through a smooth, featureless Earth, the flood (with Noah's ark visible), the modern continents (bottom), and then a global conflagration.*

an outer shell about 800 km (500 miles) thick, containing two further concentric shells and, finally, a solid central sphere. The gaps between, he said, were filled with air. The diameters of the inner shells and the central sphere corresponded to the diameters of Mercury, Venus and Mars.

Halley reckoned that the inner shell was lined with a 'Magnetical Matter', which explained the anomalies that troubled him. Gravity kept the structure intact and prevented the inner sphere from rattling around and colliding with the walls of the outer shells.

Halley went on to suppose that the empty space inside Earth was filled with living creatures. He assumed that the interior was 'replete with such Saline and Vitrolick Particles' that any gaps in the outer shell could be plugged, to avoid the ingress of water.

Journeys to the interior

Since Halley's time, others have been seduced by the theory of a hollow Earth. In 1818, the American army officer John Cleves Symmes Jr published a pamphlet in which he claimed that 'Earth is hollow and habitable within; containing a number of solid concentric spheres, one within the other, and that it is open at the poles 12 or 16 degrees.' Until his death in 1829, he lobbied for support for an expedition to explore this inner world.

A Norwegian, Olaf Jansen, claimed to have sailed through an entrance to Inner Earth at the North Pole in 1811. He described living there for two years with a race of superhumans who were 3.6 m (12 ft) tall. The Nazi leader Adolf Hitler is also thought to have believed in a hollow Earth; other senior Nazis certainly did, apparently mounting an expedition at one point. There are hollow Earth conspiracy theorists today, struggling to defend the notion against the twin onslaughts of science and common sense.

Life on the inside

While Halley envisaged unknown creatures living inside the globe of Earth, the American physician and alchemist Cyrus Reed Teed placed all of us on the inside. Teed was constantly conducting unconventional experiments and, after

The Koreshan model of Earth has the continents inscribed on the interior of a sphere looking inwards on the cosmos.

giving himself such a severe electric shock that he passed out, awoke to realize that he was the messiah. Changing his first name to Koresh, he refuted the notion that Earth revolves around the Sun and pioneered his own theory of the universe, known as the Cellular Cosmogony. By reversing the usual model of Earth covered by the astral dome, he located Earth on the inner surface of a sphere looking towards a contained central sphere that represents the heavens. He started a sect called Koreshan Unity which attracted followers who believed they could acquire immortality through the practice of celibacy and communism. Most of the sect disbanded after Teed's death in 1908.

Magnetic Earth

The idea that part of Earth's interior might be iron was proposed by the English physicist William Gilbert in 1600. Gilbert discovered Earth's magnetic field. Using a magnetized ball of iron, he found that the pattern of magnetic lines around it matched the patterns made by a moving compass needle in different places on Earth's surface. This suggested that Earth is a vast magnet and therefore must be made of iron. As more data was collected, it became apparent that Earth's magnetic field was drifting westward.

In 1692, Halley suggested that a fluid layer between Earth's crust and its core allowed the core to rotate at a different speed to the rest of the planet, which might explain the discrepancy. The magnetic field was explained in 1946 by Walter Elsasser, who stated that Earth is a geomagnetic dynamo. Moving fluid in the outer core generates electrical currents in the same way that a dynamo generates electricity.

The suspended pen of a seismograph is shaken by earthquakes or tremors and inscribes a line on the paper attached to a revolving drum.

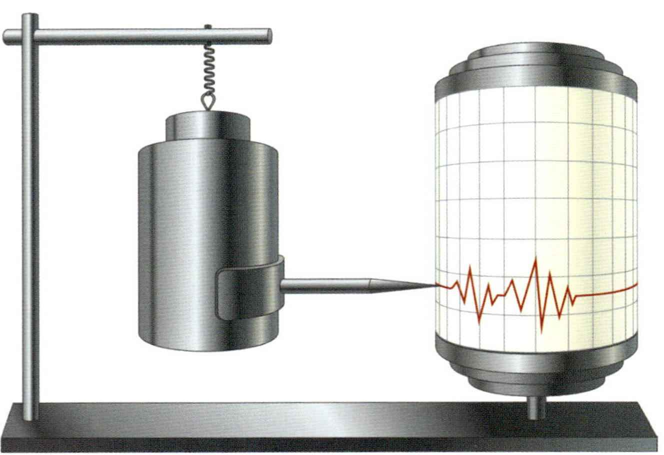

Waves in the rock

Although the evidence of volcanoes suggested a liquid layer somewhere beneath Earth's surface, the physicists of the 19th century, particularly Kelvin, theorized that if the subsurface were liquid, tidal effects produced by the Moon would tear Earth apart. Kelvin organized experiments to check for vertical movement of the surface caused by tides and concluded that Earth was 'as rigid as steel'.

It turned out, though, that Earth can carry waves of its own. Towards the end of

the 19th century, the Prussian geophysicist Emil Wiechert embarked on pioneering work to investigate how seismic waves propagate through the Earth. There are convection currents in the fluid mantle and these transmit seismic waves. Earthquakes and the shockwaves they produce were central to working out the structure of the planet. By comparing readings in different places in the hours after an earthquake, seismologists can work out how energy moves through the different types of material that make up Earth's deep structure.

In 1896, Wiechert published his theory that Earth has a rocky outer layer and an iron core, deduced from the difference between the calculated density of the planet compared with the measured density of surface rocks. His theory was confirmed in 1906 by English geologist Richard Dixon Oldham, who found that the speed at which seismic waves of energy from an earthquake travel increases with depth – but only up to a certain point. Below that, the waves slow down considerably, which indicates they are travelling through a different substance. Oldham concluded that they are slowed by a core that is much denser than its surroundings and probably made of iron.

Finding boundaries

In 1910, the Croatian seismologist Andrija Mohorovicic identified a point at which the velocity of seismic waves changes sharply, and he linked this with density. The discontinuity, now recognized as varying between 10 km (6.2 miles) under the ocean and 50 km (31 miles) under continental land, is named the Mohorovicic (or 'Moho') discontinuity, and is the division between the crust and the mantle.

Another discontinuity emerged soon after. In 1912, one of Wiechert's students, Beno Gutenberg, found that the velocity of seismic waves changes considerably at a depth of 2,900 km (1,800 miles). He identified this as the boundary between the mantle and the core (the Gutenberg discontinuity). The Earth, therefore, was conclusively shown to have three layers: the core at the centre, a thick mantle of semi-molten rock, and a hard crust.

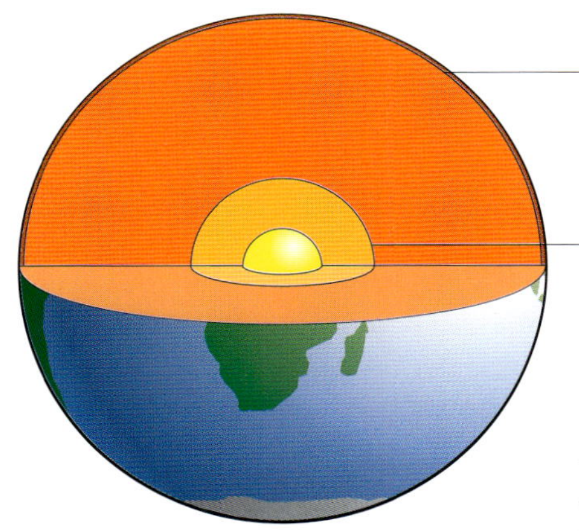

The Mohorovicic (Moho) discontinuity is the boundary between the crust and the mantle

The Gutenberg discontinuity is the boundary between the mantle and the outer core

Boundaries between layers of Earth's interior change the speed of seismic waves. From this, scientists have calculated the depths of the discontinuities.

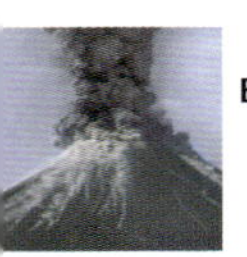

One important question remained: was the core solid or liquid? The issue was settled in 1926 by Sir Harold Jeffreys, who showed that the average rigidity of the mantle is much greater than the average rigidity of the entire planet. This had to be compensated for by an area of much lower rigidity, which could only be in the core.

Inge Lehmann deduced the existence of the inner core from her study of seismic waves.

But that wasn't the end of the story. Eleven years later, in 1937, the Danish seismologist Inge Lehmann showed that there is a solid inner core within the liquid outer core. On studying seismographic records of a major earthquake that had taken place in New Zealand in 1929, she found that some waves had travelled part way into Earth's core and then been deflected. This boundary was, she suggested, between the liquid core and a solid inner core. Her theory was confirmed in 1970 when more sensitive seismographs recorded the waves bouncing off the inner core. The boundary is known as the Lehmann discontinuity.

Geophysicists suspect there is an inner-inner core which is different again, though not substantially so. It is thought to be 1,180 km (733 miles) across and made of iron, but with a different crystalline structure from the outer part of the inner core. The inner core is thought to be growing at the rate of about a millimetre a year through the crystallization of material at the boundary with the outer core.

Layer after layer

The layers of inner and outer core, the upper and lower regions of the mantle, crust and atmosphere are the essential chemical divisions of the planet's structure. It can also be divided according to the physical behaviour of the rocks under different pressures and temperatures. The inner and outer core remain the same, but the mantle

FINAL PIECE

Earth's inner core formed relatively late in the planet's history. Estimates vary, but a 2015 study places it at 1–1.5 billion years ago. A sharp increase in the strength of Earth's magnetic field recorded in rocks has been interpreted as evidence that the core is beginning to solidify. The study also suggests that the core is cooling more slowly than previously thought.

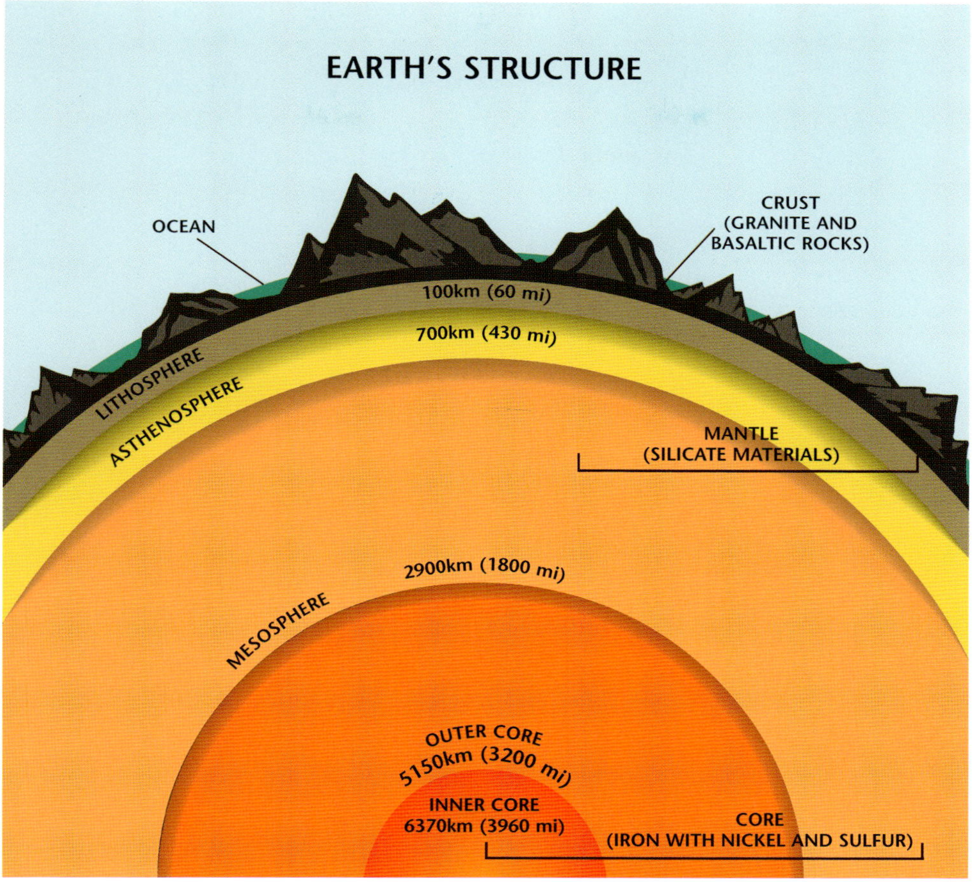

Different ways of representing the outer layers of Earth, showing the depth of the bottom of each layer.

and crust are divided into mesosphere (lower mantle), asthenosphere (most of the upper mantle) and lithosphere (the very top and mostly solid part of the mantle and the crust). The lithosphere interacts with the atmosphere (gases), hydrosphere (water), cryosphere (ice) and biosphere (living things).

Ready to go

Just a few hundred million years after the start of the solar system, Earth was ready to begin its evolution into the planet we know today. It had a single moon, an atmosphere that was largely carbon dioxide and water vapour, worldwide liquid oceans, and a rocky crust that formed islands of granite ready to grow into continents. The interior was differentiated, with a metallic core and a thick mantle of mushy, hot rock. It was quite possibly cool and rather pleasant most of the time; it might even have been ready to host its first early life forms.

CHAPTER 4

Rocks of **AGES**

'We felt necessarily carried back to a time when the schistus [rock] on which we stood was yet at the bottom of the sea, and when the sandstone before us was only beginning to be deposited, in the shape of sand or mud, from the waters of the supercontinent ocean. . . . The mind seemed to grow giddy by looking so far back into the abyss of time.'

Geologist and mathematician
John Playfair, 1788

Stone has long been analogous with stability. The rocks of Earth seem eternal and unchanging. Yet rocks do change, both mechanically and chemically, dissolving, growing, crumbling and morphing over the millennia. Relative to Earth's size, its rocky crust is as thin as the skin of an apple, yet it is here that the rest of the story of our planet has played out.

A monolith of resilient sandstone towers above the chalky desert of the Ustyurt Plateau, Kazakhstan. The landscape has been created by deposition and erosion of rock over millions of years.

All over the world

As Earth cooled, igneous rocks formed from the cooling magma; other types of rock appeared as time passed. The action of wind, weather and waves broke up some of the igneous rocks and ground them to dust or sand, which mixed with water to make clay. The dust and clay were squashed under great pressure to make the first sedimentary rocks (see box). When sedimentary or igneous rock is heated (but not entirely melted) and compressed, it can undergo physical changes which produce metamorphic rock. If the rock melts completely and reforms, it makes igneous rock.

The composition of magma is very varied, so there are different types of igneous rock. As a general rule, the

TYPES OF ROCK

Geologists group rocks into three broad types, according to how they form.

Igneous rock starts molten (as magma) and then solidifies. There are two types: extrusive and intrusive. Extrusive igneous rock is the result of magma cast out as lava through volcanic activity and hardening on the ground. These rocks cool quickly and have fine crystalline grains; if they cool very quickly, they are amorphous with no grains at all. Examples include basalt, pumice and obsidian. Intrusive igneous rock, including granite and gabbro, forms as magma hardens underground. This rock cools slowly, and grows large, differentiated crystals. It's often easy to see the separate grains of feldspar and quartz in granite, for example.

Obsidian, a black volcanic rock with no grains, at Newberry National Volcanic Monument, Oregon, USA.

continental landmasses contain a lot of granite, an igneous rock that has hardened underground, and the seabed is made largely of basalt, an igneous rock that has welled up as magma and solidified when exposed to seawater. There are around 700 types of igneous rock, and they are usually hard and heavy. Metamorphic rocks exist in pockets everywhere, and sedimentary rocks cover about 75 per cent of Earth's surface, overlying the igneous bedrock.

A multitude of uses

Our ancestors took a keen interest in the rocks they found in the ground. Some types of rock or soil could be used to make pigments for painting or dyeing cloth and pottery. Some yielded metals such as iron,

Sedimentary rock is formed in a four-stage process: the weathering of rocks breaks them into small pieces; the finer material is carried, usually by water; it is then deposited; the sediment is compacted under pressure, becoming rock. Sedimentary rock may contain organic material, such as the bodies or shells of animals, or plant matter. Examples include chalk, limestone, sandstone and clay. The first sedimentary rock contained no organic material, or perhaps just the bodies of early tiny microorganisms (see page 123).

Metamorphic rock is formed when either igneous or sedimentary rock is changed (metamorphosed) by heat and/or pressure. Buried rock is subject to great pressure and high temperatures. The chemical and physical changes produce metamorphic rock. It can be foliated, which means that it has clear layers or bands, or non-foliated. Examples of metamorphic rock include marble (from limestone), which is non-foliated, and slate (from shale), which is foliated and breaks easily into sheets.

Above: *Sandstone is a sedimentary rock laid down in clearly visible layers.*

Right: *Marble is a metamorphic rock which is common in Europe. Here it lines the shoreline on the island of Thassos, Greece.*

copper or gold. Others were extremely hard, and useful for building. Some flaked readily and could be fashioned into tools. It's no surprise that we find attempts to classify and describe rocks and minerals among the earliest proto-scientific writing.

Ochre sand in Roussillon, France. Ochre, used as a pigment, was mined here between the 18th and 20th centuries.

Greeks and gems

The first person known to have studied and described different types of rocks and minerals was the Greek philosopher Theophrastus (*c.*371–287 BC). Born on the island of Lesbos, he moved to Athens to study at the Academy founded by Plato. When Plato died, Aristotle took over as head of the Academy and, after Aristotle fled Athens, Theophrastus ran the school for 36 years. Best known for his work on botany, Theophrastus also wrote *On Stones*, a treatise on rocks and gems, and a lost work entitled *On Mining*.

Theophrastus accepted the formulation that Aristotle set out in his treatise *Meteorologica*, that all earthly substances are composed of four elements (earth, water, air and fire) with combined properties of heat, cold, dryness and dampness. Aristotle believed metals were the result of congealing moist exhalations from the earth, and minerals were the product of dry, gaseous exhalations. This idea of a contrast between wet and dry sources of mineralogical substances would survive for around 2,000 years, surfacing in a different form in the 18th century (see pages 76–8).

On Stones describes the appearance, uses and physical properties of stones, earths and minerals. The discussion of origins is largely restricted to their mining or formation from rock and water, although it notes that one gem, lyngourion, is thought to have been crystallized from the urine of a lynx (see box at the bottom of the facing page). Most other ancient and medieval texts on stones explored their mythical origins extensively and described the supposed medicinal properties of certain minerals.

'Of the substances formed in the ground, some are made of water and some of earth. The metals obtained by mining, such as silver, gold and so on, come from water; from earth come stones, including the more precious kinds, and also the types of earth that are unusual because of their colour, smoothness, density, or any other quality.'

Theophrastus, 3rd century BC, *On Stones*

Left: *Theophrastus had wide-ranging interests – geology was just one area in which he worked.*

Theophrastus notes how stones behave when heated or 'burnt', and which ones attract other materials. (We now know that this means they are magnetic or can be induced to hold static electricity.) He explores whether they are hard or will crumble, and how they are mined, used and valued, and notes that some stones are found within others. He mentions one pebble, composed of two different types of stone, which he concludes 'had not yet entirely changed from the watery state'. *On Stones* remained the most rational work on mineralogy for nearly 2,000 years.

'Lyngourion . . . is very hard, like real stone. It has the power of attraction, just as amber has, and some say that it not only attracts straws and bits of wood, but also copper and iron, if the pieces are thin, as Diokles used to explain. It is cold and very transparent, and it is better when it comes from wild animals rather than tame ones and from males rather than females; for there is a difference in their food, in the exercise they take or fail to take, and in general in the nature of their bodies, so that one is drier and the other more moist. Those who are experienced find the stone by digging it up; for when the animal makes water, it conceals this by heaping earth on top.'

Theophrastus, 3rd century BC, *On Stones*

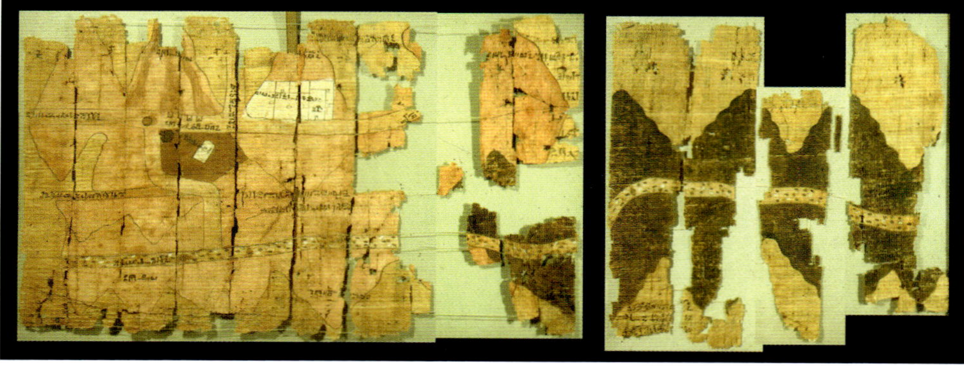

The world's earliest geological map was made 3,000 years ago. It shows where to find gold in Egypt.

Minerals and miners

Mining is one of the first ways in which humans made permanent changes to the Earth. The ability to extract and smelt metals drove the making of tools and weapons in the Iron and Bronze Ages, and would later be deployed in the manufacturing industries of the modern world.

With the advent of movable type and increasing literacy, information about mining and different types of rock spread more widely. The German scholar and scientist Georgius Agricola (1494–1555) wrote extensively about mining and geology, and established geology as a discipline. His pioneering work *On the Nature of Metals* classified minerals, earths, stones and metals on the basis of their physical properties.

Although Agricola occasionally noted the resemblance of some fossils to organisms, he stopped short of suggesting that they were their organic remains. (At the time, the word 'fossil' literally meant 'thing dug up' and did not necessarily describe something of organic origin.) The recognition that some fossils were the relics of once-living things would soon add a new dimension to geology (see page 153).

AGRICOLA AND HERBERT HOOVER

Agricola wrote in Latin, as was usual at the time. His *On the Nature of Metals* was first translated in 1912 by Herbert Hoover, a mining engineer and later 31st President of the United States. Hoover's wife Lou Henry, a geologist and Latin scholar, also worked on the translation. The task took them five years to complete.

Neptunism and Plutonism

Theophrastus' statement that materials found in the ground came from either water or earth prefigured an 18th-century debate about the origin of stones. There were two opposing theories: Neptunism (named for the Greek god of the sea) held

that rocks formed originally from water; Plutonism (named for the Greek god of the underworld) held that rock is formed underground through the action of heat.

Volcanoes and fossils

The Italian abbot Anton Moro was a geologist and naturalist who studied volcanic islands. Around 1750, he determined that the volcanic rocks he was studying had come from inside Earth and solidified after emerging (Plutonism). Moro was the first to distinguish between volcanic rocks that had formed the islands and sedimentary rocks that had been laid down afterwards and contained fossils. In his book *Of crustaceans and other marine bodies found on mountains*, he wrote that fossils of sea creatures found in mountain rocks were evidence not of Noah's flood, but of rocks that had once been buried under the sea.

The German professor of mineralogy at Freiburg, Abraham Gottlob Werner (1749–1817) believed that early Earth had accumulated from cosmic matter and initially taken the form of an ocean rich in dissolved elements. Rocks formed as minerals crystallized and precipitated out of the ocean (Neptunism). According to his model, precipitation happened in a strict sequence, with the oldest, hardest rocks such as granite and gneiss forming first, then basalt and, finally, sedimentary rocks such as limestone. After these rocks had formed, the sea level dropped and some of them were exposed. The rocks immediately began to erode, so began to form the most recent sedimentary rocks, such as sandstone.

Basalt columns in Zlaty vrch, Czech Republic, are exactly what Moro was talking about: rock that had flooded, molten, from a volcano and then hardened.

But then, in 1806, while studying the geology of the Tyrol, a mountainous region of northern Italy and Austria, the Italian mining engineer Count Giuseppe Marzari-Pencati found granite rocks lying on top of marble. Neptunism deemed this to be impossible, as granite was considered the earliest rock. Marzari-Pencati published his findings in 1820 to the consternation of the Neptunists, one of whom, Leopold von Buch, suggested that a landslide had mixed up the order of the rocks. But his argument did not stand up under scrutiny. Geologists flocked to the region to investigate, including one of the greatest naturalists of the century, Alexander von Humboldt. Neptunism would suffer further with the work of Charles Lyell in the 1830s.

People generally use the rocks that are found in their locality. This statue of the Virgin Mary in Innsbruck, Austria, is made of marble, a rock common in the Alps.

SEDIMENTARY ROCK

Geologists recognize four categories of sedimentary rock:

Clastic sedimentary rocks are made of small particles of rock that have been compacted, then cemented together by silicates. They are divided into three groups by particle size, which relate to gravel, sand or mud. The latter, with such small particles, appears not to be particulate. Silt lies between the sand and mud classes, but is often grouped with mud.

Biochemical sedimentary rocks are formed from living things. Principal examples are limestone, made from the calcium-rich bones and shells of animals; coal, made from wood; and chert, made from the remains of organisms that use silicon to build their skeletons (microorganisms such as diatoms and radiolaria).

Chemical sedimentary rocks develop when minerals precipitate from a super-saturated solution – rock salt, for example.

Rocks of life and death

In reality, both water and fire contribute to making the different rocks that Earth offers. Heat melts and changes rock; meanwhile, water carries dissolved minerals though rocks and deposits them in veins, or dissolves substances from rocks, or carries sediment to a place where it can build up and eventually form sedimentary rock. And there is another force at work – the biological sequence of life and death that has transformed the bodies of organisms large and small over billions of years. As we shall see in Chapter 7, life has wrought immense changes on Earth over the last four billion years, some of which include shaping the very substance of the planet and creating certain types of rock – limestone and coal, in particular.

The mountain of reefs

The Dolomites are a range of mountains in Italy made principally of a sedimentary rock composed of a mineral called dolomite, a carbonate of calcium and magnesium ($CaMg(CO_3)_2$). The rock was discovered in 1791 by a French naturalist, the extravagantly named Dieudonné-Sylvain-Guy-Tancrede de Galet de Dolomieu, when walking in the Alps. He noted that unlike regular limestone it contained crystals which did not respond to acid. The stone eventually gave the Dolomites their name – they had previously been called simply 'the pale mountains'. Fossils found there revealed that these mountains had once been under the sea, but little was known of the seabed and the processes which might have taken place there.

The true nature of the Dolomites was discovered accidentally and dramatically in 1770 by Captain James Cook when he ran his ship HMS *Endeavour* aground on the Great Barrier Reef off the Australian coast. Reefs had been described in a 1704 paper submitted to the Royal Society in London:

'Other' sedimentary rocks include sediments laid down by pyroclastic flows following volcanic eruptions. (A pyroclastic flow is a fast-moving cloud of volcanic matter that pours over the landscape, depositing ash.)

People have exploited the tendency of flint (a type of chert) to flake in order to make tools and weapons.

'There are big banks of this coral, it is porous and so hard or yet as smooth as the upright, which grows in small branches. If, of which we speak, is fully grown, others grow in between it, where still others will grow, until the whole structure is as hard as a rock.'

A German naturalist, Georg Forster, who sailed with Cook in 1772–5, studied the corals of atolls and volcanic islands. He discovered that although a reef might rise 300–600 m (984–1,968 ft) above the seabed, living coral was found only in the top few metres. Forster suggested that either the reef was growing upwards from the seabed, and the top was being eroded, creating a flat atoll, or volcanic activity was pushing the coral to the surface.

In the 19th century, the naturalist Charles Darwin began to explain the connection between atolls and coral. He realized that coral-forming animals needed sunlight, which explained why the coral is not built in the deep sea. He surmised that corals begin to colonize on volcanic summits that are sinking but are still close to the surface of the ocean. As the volcanoes slowly subside, the coral continues to grow near the surface by building on top of existing coral, its construction keeping pace with the sinking volcano. Darwin suggested that three related features – volcanic islands fringed by a reef, islands with barrier reefs, and atolls – show three different stages of the same process.

In 1868, the German zoologist Carl Semper found all three reef types coexisting on the island of Palau in the Pacific. Ten years later, the oceanographer John Murray suggested that coral don't restrict themselves to volcanic mounts but will colonize any suitable underwater structure. The American geologist Alexander Agassiz supported this view.

Eventually the Dolomites were recognized as the remnants of reefs that had once grown in a warm sea. We now know that they were built in the Permian period, more than 250 million years ago, and ended up in the middle of Europe

Seen from above, an atoll in Queensland, Australia clearly preserves the shape of a volcanic crater.

when the African plate and the Eurasian plate collided, pushing up mountains from the seabed. In the Dolomites, there are areas of mountain which were made almost entirely from or by living creatures.

The Dolomites are evidence of prehistoric coral reefs, now stranded far inland.

LAYER AFTER LAYER

There are many places where it is easy to see that rock has been laid down in layers, or strata. The Chinese Shen Kuo, Persian Ibn Sena and Italian Leonardo da Vinci all made discoveries about the deposition and erosion of rocks, but it was the Danish geologist Steno who set out the following principles of stratigraphy in 1669:

- Layers of rock are laid down in order, so the lowest layer is the oldest.
- The lower layer has become solid before the next layer is laid down.
- Anything above the stratum being laid down must be fluid (liquid or gas).
- The edges of a stratum must be bounded by some other solid, or the stratum must extend around the entire world.
- If anything cuts across a stratum, it must have formed after the stratum.

Steno also wrote about the inclusion of one solid within another, for example, crystals, incrustations, fossils, veins within rocks and strata. He proposed that fossils were the remains of ancient living organisms and observed that the organism which solidifies lends its shape to the subsequent layers. He therefore reasoned that fossils cannot form within solid rock, as had previously been claimed. In strata, the shape of the lowest layers determines those laid down on top. Crystals and veins running through rocks are often deformed by the need to fit to the gaps or pressures of the existing rock.

While Steno's precepts are still pertinent, they don't explain how rocks are formed. The next important step was taken by James Hutton at the end of the 18th century and concerned not just the formation of rocks, but the way in which they are worn away.

Eaten away

Erosion is the wearing away and removal of rock, usually by the action of wind, water or ice; the removed material is carried away and deposited somewhere else as a sediment. The process of erosion begins as soon as any rock is exposed to water or weather, so it occurred very early in the story of Earth – as soon as there was a combination of solid rock, water and atmosphere.

The work of wind and water

Water erodes rock by washing over it, often carrying grains or stones that wear away the rock by constant abrasion. Frozen water, in the form of glaciers, can do this very powerfully. A glacier is a slow-flowing mass

Leonardo's Virgin of the Rocks *(Paris version) includes accurately observed geological features. The grotto is made of weathered sandstone dissected by a layer of harder rock. The vertical rock above the Virgin's head is diabase, an igneous rock that intruded into the sandstone when it was molten and spread, forming a thick band (or sill). A horizontal crack above the diabase marks the start of the next layer of deposited sandstone. The sandstone is shown weathered and rounded, particularly in the roof, whereas the harder diabase is resilient and remains angular. The sandstone in the foreground has clear layers.*

***Above:** The Grand Canyon has been cut through rock by the Colorado River over the course of 35 million years. The rocks of the cliffs have been further eroded by wind and weather.*

***Right:** Erosion by flowing water produces a V-shaped valley, but erosion by flowing ice, in the form of a glacier, produces a wider, U-shaped valley like the one in the foreground. Small tributary glaciers don't erode as effectively as large glaciers, so when the ice recedes they can leave a 'hanging valley' which stops short of the main valley floor.*

of ice; it is immensely heavy and can carry everything from sand grains to boulders. Wind carries fine dust, or even quite large grains of sand that grate away the surface of rock. Softer rocks are carved away more quickly, which can lead to interesting shapes where different strata are exposed.

Left: *Hoodoos ('fairy chimneys') in Cappadocia have been formed by the wind eroding soft rock and leaving in place harder volcanic rock. Here, the hard rock forms a cap on top of a column of softer rock.*

Below left: *The most powerful erosion agent is ice. A glacier is slow-moving but heavy and often carries loose stones and boulders with it. If these are on the underside of the glacier, they scrape grooves into the rock, as here.*

The work of weather

Weathering reduces rock but does not include an agent of movement to carry away the removed particles. Geologists recognize three types of weathering: chemical, physical and biological.

Chemical weathering is produced by the action of rain. Often slightly acidic, rain dissolves carbonate rock. As the carbonate dissolves, other grains in the rock are freed to fall away.

Physical weathering is the result of changing temperature. Where water seeps into rocks and freezes, the expanding ice cracks the rock. Repeating the freeze-thaw cycle can break up even large bodies of rock. In deserts, rocks expand and contract as they heat up and cool down, the stress eventually leading to horizontal cracking.

Biological weathering is caused by living things. For example, a tree can send roots down into the gaps between rocks and crack the rocks as it grows. Lichen, algae and bacteria produce chemicals which dissolve the surface of the rock. Shellfish such as the piddock dissolve or scrape holes in rocks to establish a home.

Recycling rock

James Hutton was the first to suggest that Earth is the product of eons of physical change. He recognized that the processes which shape the Earth must be very slow, and our planet must be much older than was generally believed. By studying the rock formations in Scotland, where he lived and farmed, he observed how the strata are organized and rocks are disrupted, eroded and deposited in an ongoing process.

Geologist James Hutton with his hammer, examining exposed rock strata.

Hutton observed rocks along the Scottish coast in Berwickshire, where different strata are exposed. He concluded that neither the Plutonists nor the Neptunists had correctly answered the question of how rocks form. He saw that sedimentary rocks are laid down by water, but that igneous rocks are quite distinct from these and must have evolved differently. He recorded the gradual processes of weathering and erosion. Where he saw that strata were disrupted, not lying in neat horizontal layers as Steno dictated but angled and even folded, Hutton realized that some powerful force had shifted the land after the rock had been laid down.

A local landmark, Siccar Point, provided an abundance of evidence. Here vertical layers of grey shale lie beneath horizontal layers of red sandstone. Hutton realized that the gray shale had been deposited, then

JAMES HUTTON, 1726–97

Often referred to as the 'father of geology', James Hutton was born in Edinburgh, Scotland, one of five children. When he was only three years old, his father died. At the age of 14, Hutton went to Edinburgh University, first studying classics and becoming apprenticed to a lawyer, then turning his attention to medicine and chemistry. After studying in France and the Netherlands, he returned to Scotland where he had inherited two farms. He developed an interest in the geology of the area and in how water and weather affected the land. In 1767, he moved back to Edinburgh and pursued his interests in geology, fossils and experimental science, forging friendships with some of the leading intellects of the day including the chemist and physician Joseph Black, the economist Adam Smith and the philosopher David Hume.

In 1785, Hutton presented a landmark paper to the Royal Society of Edinburgh (which he had co-founded in 1783) explaining his theory that Earth had been shaped by powerful geological forces over the course of many more years than was generally believed. He described a nearby site, Siccar Point, where layers of rock testified to his account of uplift and deformation.

FAULTS AND FOLDS

There are two main types of disruption in rock strata:

Folding occurs when the layers are literally folded or wrinkled up, so that they form waves.

Faulting occurs when there is a break across the layers and a whole chunk of rock slips downwards or is tilted.

Above: *Folding, the result of pressure from below, has produced wave-like strata in this deposit of gneiss.*

Left: *These rocks in the Seppap Gorge, Morocco, show clear faulting: the rocks to the left have slipped downwards so the layers no longer line up left to right.*

some force had lifted and tilted the entire layer of rock, which had been eroded and covered by an ocean. The red sandstone had been laid down on top of the upended layers. Further changes led to the ocean disappearing and the entire formation lying inland in Scotland.

Hutton concluded that rocks are created and destroyed in a 'great geological cycle' which must have been underway for far too long for Earth to be only 6,000 years old, as many people believed. With stunning insight, he wrote that the rocks we see today are made from 'materials furnished from the ruins of former continents'. In this cycle, rocks and soil are washed from the land into the sea, where they are compacted into bedrock. Volcanic forces then push them upwards to the surface, where they are eventually worn away into sediment again.

Hutton recognized that the process is driven by heat from within the Earth operating over extremely long timescales. He suggested that the existence of volcanoes

'The imagination was first fatigued and overpowered by endeavouring to conceive the immensity of time required for the annihilation of whole continents by so insensible a process.'
Scottish geologist Charles Lyell, writing about Hutton's theories

The steeply tilted grey rocks lower down formed first. Over a period of 65 million years, while no new rock was deposited, faulting, folding, uplift and erosion changed the base layer. The younger, red rocks were laid down on top of them after the ocean had covered the original layer.

and hot springs was connected with this subterranean heat, and theorized that high temperature and pressure could have physical and chemical effects. The physical effects included the expansion of Earth's crust and pressures that pushed up mountains, and folded, tilted and deformed rock. The chemical effects could form granite, basalt and add the veins of different minerals that run through many rocks. Although Hutton's account contained gaps which could only be filled with a 20th-century understanding of tectonics, it was an astonishing feat of deduction.

UNIFORMITARIANISM

Hutton formulated the notion of 'uniformitarianism', which states that the same processes operate now as have always operated, and will continue to do so in the future. Although geological processes are so slow that we don't notice them happening, they are still progressing. If we can measure, for example, the rate at which sedimentary rock is formed, we can calculate how long any particular seam of rock has taken to form since the process does not change in nature or speed from one eon to the next. Most importantly, because processes were not different in the past, 'the past is the key to the future'.

CHAPTER 5

Active **EARTH**

'Water, Fire; Fire, Water; mutually, as it were, cherish one another; and by a certain unanimous consent, conspire to the Conservation of the Geocosm, or Terrestrial World.'

Athanasius Kircher, 1665

The cycle of rock formation that was unravelled by James Hutton and later geologists relates only to the crust, the top surface of Earth. But this process signifies something much larger – the eternal turmoil within the planet.

Volcanic eruptions are part of the process of recycling rock on our planet, with magma from within emerging as lava which will harden into new surface rock.

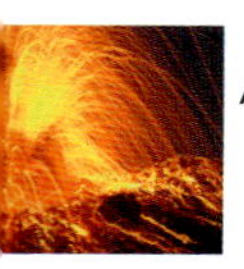

Heat put to work

Hutton's uniformitarianism was a challenge to the prevailing catastrophist model. The latter assumed Earth was shaped by a series of sudden events or catastrophes which produced change over a short timescale. It held that the past included at least one great global flood, which was sometimes linked with Noah. Other local catastrophes, such as eruptions, earthquakes and tsunami, could be witnessed more regularly and recently. In 1755, a catastrophic earthquake beneath the Atlantic Ocean had destroyed the city of Lisbon in Portugal and was fresh in the minds of European thinkers.

The kinds of geological change that have been going on for millions, even billions, of years are the result of cataclysmic events such as earthquakes and volcanic eruptions and the very gradual changes that Hutton described. By referring to subterranean heat and pressure causing uplift, folding, tilting and faults, Hutton was noting outcomes

THE LISBON EARTHQUAKE

In 1755, an earthquake just off the coast of Portugal caused devastating destruction in areas of Portugal, Spain and Morocco. Tremors were felt as far away as Greenland. The earthquake was followed by a tsunami that was 20 m (66 ft) tall and possibly reached as far as Brazil. The earthquake occurred on All Saints' Day (1 November) when many homes and churches were burning candles. Falling candles started a fire that razed most of the city of Lisbon.

The death toll is not known exactly, but the combination of earthquake, tsunami and fire probably killed between a seventh and a half of the population of Lisbon and a third of the population of the Spanish city of Cadiz. The impact on Europe was dramatic, casting into doubt the idea of a just God ordering human affairs for the best, and the entire notion of a stable Earth. One outcome of the earthquake was the emergence of seismology as a western science, born of the struggle to understand the catastrophe.

Falling buildings following the 1755 earthquake and floods from the subsequent tsunami wreaked havoc in Lisbon.

which were sometimes the fossilized aftermath of past cataclysms and sometimes the result of slow, relentless change.

Moving lands

As Earth cooled and the crust solidified and grew into the first continents, the interior continued to move. Convective currents in the magma brought hot material towards the surface as subducting colder material descended. The results of this movement became apparent long before their cause was known.

AN EARLY PUZZLE

From a quick glance at a world map, it is clear that the outlines of Africa and the Americas fit together well, like pieces of a jigsaw puzzle. In 1596, the Flemish mapmaker Abraham Ortelius commented on this and conjectured that the Americas had been 'torn away from Europe and Africa . . . by earthquakes and floods'.

Later, people discovered other contiguities between the continents. In 1858, the French geographer Antonio Snider-Pellegrini pointed to similar fossils found on both continents. There are several types of animal fossil found in bands that cross Africa and South America. Fossils of the plant *Glossopteris* are found across South America, Africa, India, Antarctica and Australia, a fact which led the Anglo-Germanic geologist Eduard Seuss to suggest that these landmasses had once been joined together in a continent which he named 'Gondwanaland'.

But how could such a large gap have opened between the continents? In 1912, a German meteorologist and geophysicist,

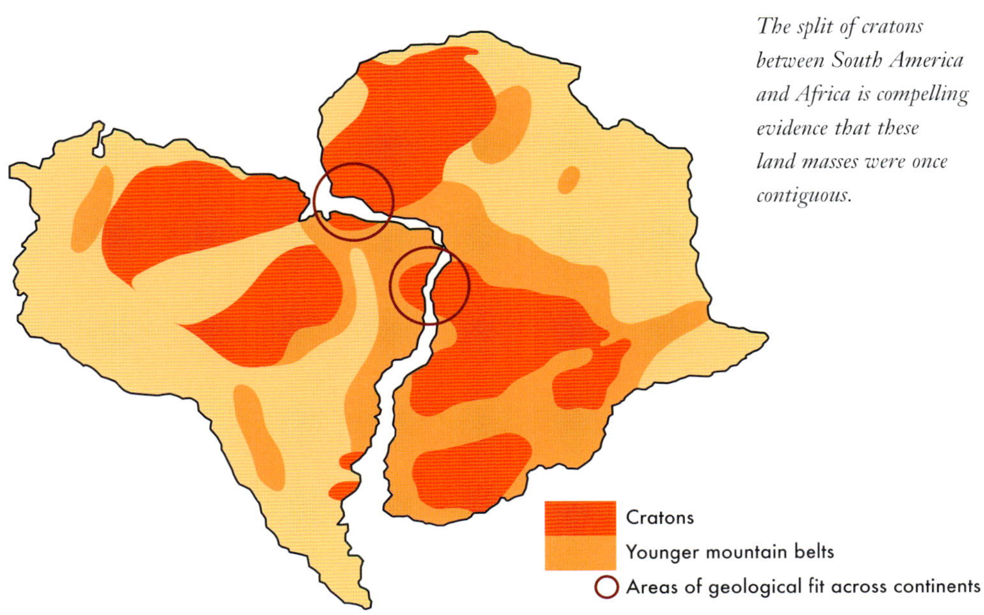

The split of cratons between South America and Africa is compelling evidence that these land masses were once contiguous.

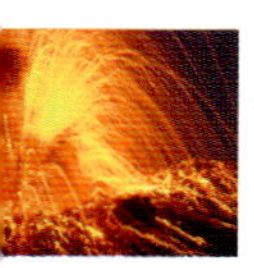

Alfred Lothar Wegener, suggested his theory of continental drift in *The Origin of Continents and Oceans*. His interest in the alignment of the landmasses had begun in 1911, when he read a paper about similar plant and animal fossils found on each side of the Atlantic. He set about finding other fossils which matched in these divided lands and discovered there were a lot of them. The prevailing explanation was that there had been land bridges between now separate landmasses, but Wegener found this unsatisfactory.

He extended his search beyond fossils and found matching geological features separated by oceans. For example, he discovered that the Appalachian Mountains of eastern North America geologically match the Scottish Highlands, and the distinctive rock strata of the Santa Catarina system in Brazil match those of the Karroo in South Africa. Some cratons now known to be two billion years old are split across widely separated continents. Wegener discovered that the fossils found in some areas, such as tropical plants found in Antarctica, are of organisms completely unsuited to the current climate of that region.

> *'Wegener's hypothesis in general is of the footloose type, in that it takes considerable liberty with our globe, and is less bound by restrictions or tied down by awkward, ugly facts than most of its rival theories.'*
>
> Rollin T. Chamberlin

His conclusion was that around 300 million years ago all the continents were joined in a single large landmass, Pangea. This split into the current continents, which drifted apart. But Wegener could not suggest any convincing mechanism for moving the continents. His notion that they had ploughed through Earth's crust or been moved by tidal forces was ridiculed.

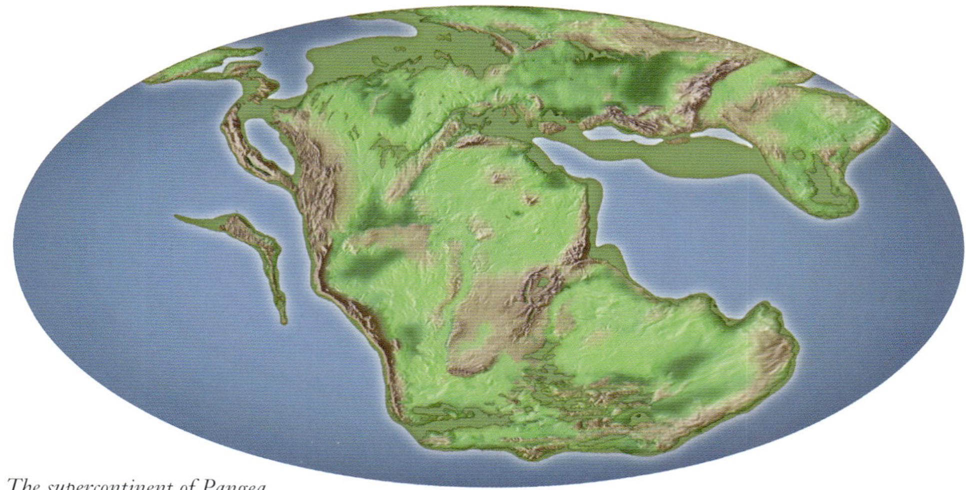

The supercontinent of Pangea, 200 million years ago.

ALFRED LOTHAR WEGENER, 1880–1930

Alfred Wegener was born in Berlin, Germany, the son of a clergyman and the youngest of five children. He gained a PhD in astronomy in 1905, but was also drawn to geophysics and meteorology. In 1906, he joined an expedition to Greenland to study polar air circulation, the first of four visits. Wegener's trips to Greenland were nothing if not eventful. Three men died on his first expedition, which established the first weather station in Greenland and mapped the last stretch of unmapped coast. During his second expedition, in 1912–13, a calving glacier almost killed the team. Wegener and Johan Koch (who was injured in the accident) were the first people to overwinter on the north Greenland ice and made the first northern crossing. They ran out of food in difficult terrain near the end of their journey, and had already eaten their last dog and pony when they were found and rescued by a clergyman visiting a remote congregation.

On his return to Germany, Wegener took a post at Marburg University, which is where his interest in continental drift took root. His academic career was briefly interrupted by World War I, but he was invalided out of active service and assigned to meteorological work. He worked on tornadoes, but continued to refine and promote his continental drift theory. He made two more trips to Greenland. On the final trip, Wegener and a companion set out to move supplies to the camp on the west coast in bad weather, but died en route.

Geological evidence supporting the idea of continental drift only began to emerge 30 years after his death.

Sonar discoveries

During World War II, the American geologist Harry Hammond Hess was commanding an attack transport ship. Interested in the profile of the seabed, he left the sonar used to track submarines turned on constantly as his ship sailed the North Pacific to map it. Sonar works by bouncing sound waves off an object and using the time the echo takes to return to calculate the distance from it. Hess expected the ocean floor to be flat, but found a landscape of ridges, canyons and mountains, as on dry land. Further work revealed the Mid-Atlantic Ridge, with mountains rising sometimes above sea level to form islands (such as the Azores and St Helena). The deepest parts of Earth's oceans are trenches close to the continental landmasses. The Marianas Trench off the coast of Japan plunges to more than 11 km (6.8 miles).

To explain his findings, Hess proposed that oceans grow from the middle. In 1962, in *The History of Ocean Basins*, he described a

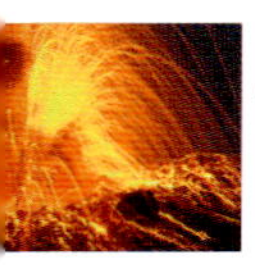

mechanism that has basalt lava oozing from the seabed at the ridges, where the Earth's crust is very thin, and piling up next to them. The hot new rock has a greater volume than the nearby cold rock, which explains the height of the ridges. As the new rock cools, it sinks. The seabed, Hess reasoned, moves constantly from the middle of the ocean, where it is created, to the edges.

The deep troughs near the continental landmasses are where the rock of the ocean floor is destroyed and recycled. It is pulled beneath the continental crust and melted back into the mantle at subduction zones. Subduction releases water that has been carried down with the melting seabed and makes the magma more liquid, leading to frequent eruptions from volcanoes that are building just beyond the subduction zone. The 'Ring of Fire' around the rim of the Pacific Ocean has 452 active volcanoes.

There are two immediately obvious consequences of Hess's discoveries: one is that the rock of the seabed is generally newer than that of continental land and the other is that the landmasses can move slowly around the planet – the continental drift that Wegener proposed but was unable to explain. Hess's explanation was neat, but without geological evidence it was unlikely to gain traction. Luckily, geological support came along a year after the publication of Hess's book.

Stripes under the sea

Two British geologists, Frederick Vine and Drummond Matthews, had been studying magnetized stripes in the rock of the ocean floor. These stripes occur because the Earth's magnetic field switches direction at intervals (called geomagnetic reversal), so the magnetic north pole moves to the south pole. Over the last 20 million years, this has happened on average every 200,000–300,000 years – though the last switch was 780,000 years ago. (A switch probably takes thousands of years, so even though a switch looks overdue, North Korea won't become South Korea overnight, nor North America change to South America.)

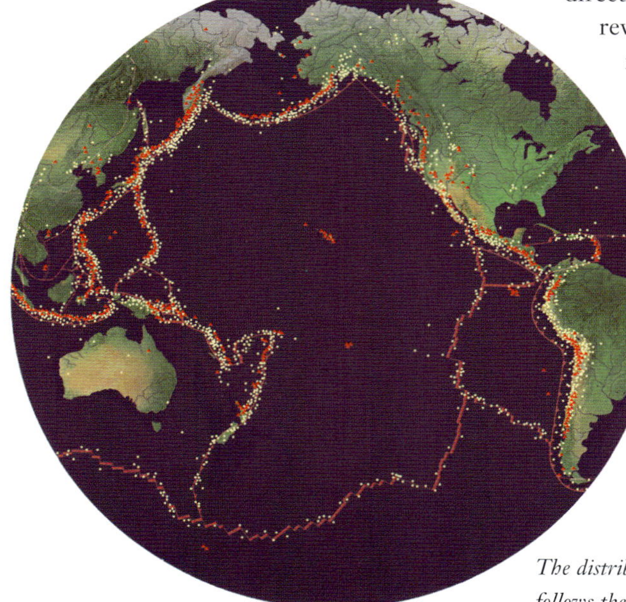

The distribution of volcanoes around the Pacific follows the outlines of the crustal plates.

As new basalt rises mid-ocean, the existing sea floor moves apart, preserving a record of Earth's magnetic history. It is shown here as alternating pale and dark blue stripes representing switching polarities.

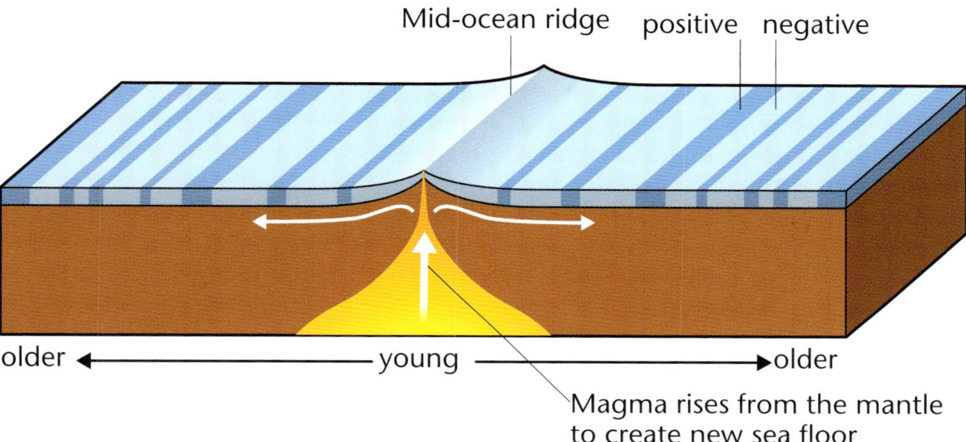

Basalt as it emerges from the mantle is a thick, sluggish liquid. It contains an iron oxide called magnetite, which is highly magnetic. The magnetite lines up in the basalt so that it is oriented north–south and as the basalt cools the magnetite is frozen in position. This means it preserves a record of Earth's geomagnetic orientation at the time it was laid down. The stripes of differently oriented magnetite in the rock of the seabed can be read to reveal a history of geomagnetic reversals over time. Vine and Drummond realized that by looking at the patterns of magnetism around the ridges they could test Hess's theory. It turned out that the patterns of reversing magnetism are symmetrical around the mid-ocean ridges, suggesting that the emerging basalt splits, with half going to each side of the ridge, and when frozen in place it shows the same pattern. Working from the ridge towards the edges of the ocean, a matching history of magnetic reversals emerges.

Hot spots

While the theory of continental drift gained momentum, some questions remained unanswered. An obvious one was the puzzle over why some volcanoes and earthquakes occur far from the mid-ocean and continental boundary flashpoints.

In 1963, the Canadian geophysicist John Tuzo-Wilson revealed that the Earth's crust moves over stationary 'hot spots' in the mantle below. These hot spots represent upwellings of magma which burst through the crust and build over time into large, low shield volcanoes. As the crust slowly moves over the mantle, the area above the hot spot changes on geological timescales. Tuzo-Wilson's insight explained the existence of chains of volcanic mountains such as those which form the Hawaiian islands today.

Two years later, Tuzo-Wilson solved another part of the puzzle. Up until this point, only two types of boundaries – destructive and constructive – had been

recognized. Destructive (or convergent) boundaries occur at the junction between oceanic and continental crust, where the moving oceanic crust is pushed down into the mantle and destroyed by subduction. Constructive (or divergent) boundaries occur at rifts, such as those in mid-ocean, where the plates are moving apart and magma wells up from below to form new rock. Tuzo-Wilson proposed a new type, called a conservative boundary (or a transform fault). At this point, parallel slabs of crust slide alongside each other in opposite directions, with nothing being destroyed or created. Conservative boundaries, such as the San Andreas Fault in California, are often the site of earthquakes as tension builds up between slabs of crust that catch on each other, and is suddenly released when they finally move.

Moving along

Although the pieces of the puzzle seemed to fit together well, the cause of movement in the Earth's crust was still not understood. Then, in 1966, the British geophysicist Dan Mckenzie applied thermodynamics to the problem. He suggested that the mantle

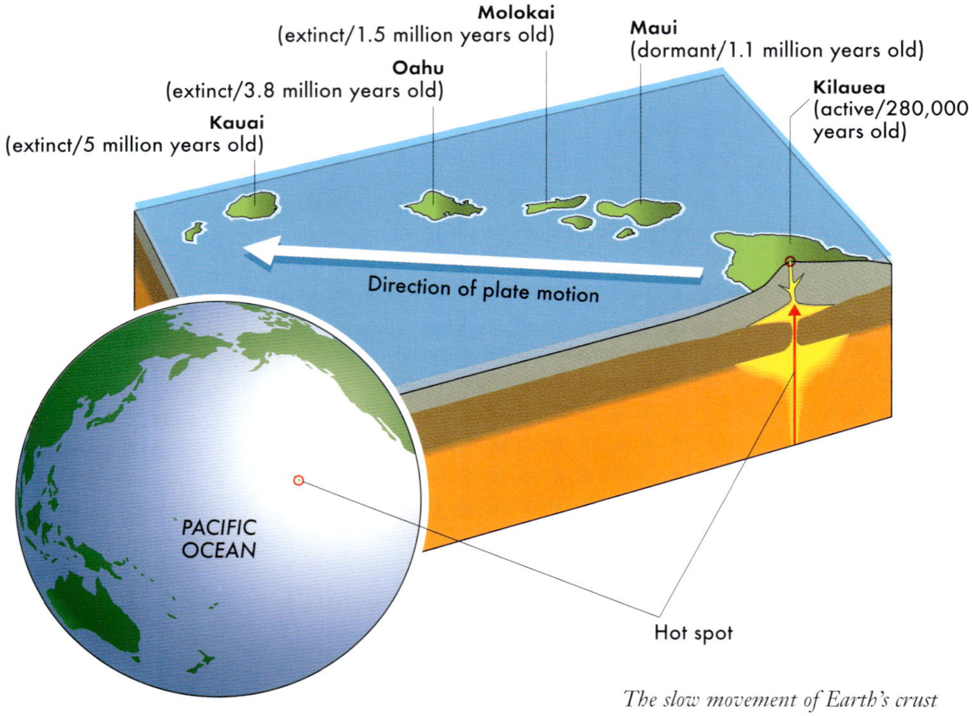

The slow movement of Earth's crust over a hot spot creates a chain of volcanoes, as demonstrated here in the Hawaiian islands of the Pacific Ocean.

has two layers, each of which is in motion in different ways. The crust floats on top of the upper mantle and moves with it. In 1967, the American geophysicist W. Jason Morgan proposed a model with 12 chunks of crust. Known as tectonic plates, these chunks move relative to one another. In 1968, the French geologist Xavier le Pichon published a complete model, with six tectonic plates.

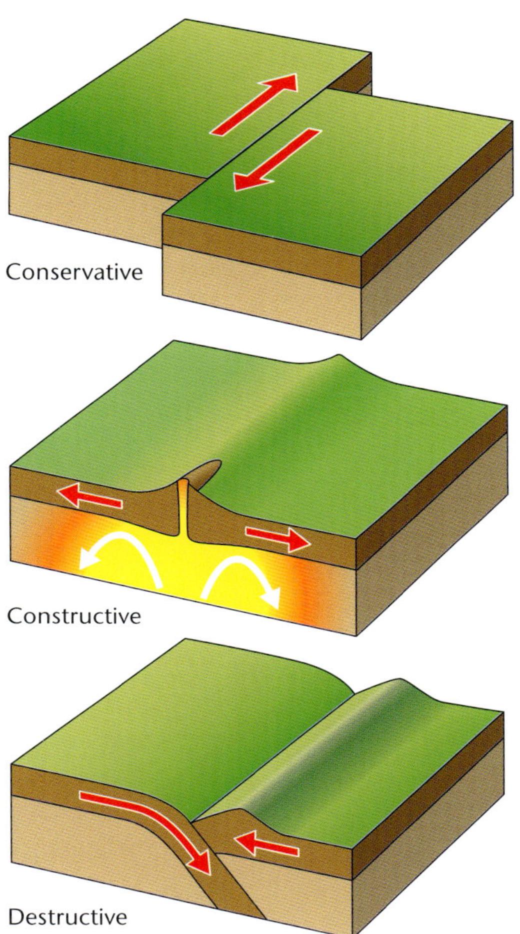

The movement of tectonic plates at conservative, constructive and destructive plate boundaries.

Plate tectonics: a scientific revolution

During the 1960s, Alfred Wegener's derided theory of continental drift was rehabilitated as plate tectonics theory. It was one of the most important developments in Earth science, explaining with one model the formation and behaviour of Earth's crust, events such as earthquakes and volcanic eruptions, mountain-building, the illogical-seeming distribution of rocks and fossils, the mid-ocean ridges and location of volcanoes. The acceptance of plate tectonics is generally dated to 1965, when Edward Bullard showed the best fit of land on the east and west sides of the Atlantic if the ocean were closed up ('Bullard's Fit').

The current model has Earth's crust divided into tectonic plates which are moved slowly but not uniformly by convective currents in the magma below. The plates jostle, collide and scrape against one another and occasionally change shape when they break apart or stick together.

Breaking plates

Although the plate tectonics model is now widely accepted, it's difficult to work out what happened in the early days of Earth, or when current tectonic activity began. As the oceanic crust is constantly recycled, the oldest parts (near the shore) are only about

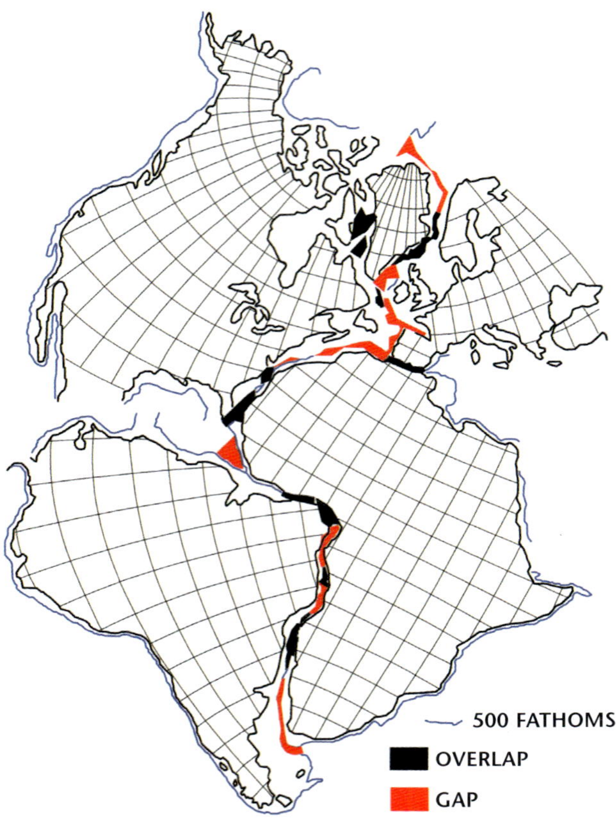

Edward Bullard's computer-generated 'best fit' for a supercontinent created by a closed Atlantic Ocean.

200 million years old. The oldest exposed rocks on land of any size are nearly four billion years old. Geologists still don't know when the first continents formed or when the crust broke into separate tectonic plates. Modelling suggests that plates might have formed three billion years ago, but they are possibly much newer. It's likely that the crust initially formed one single plate which wrapped around the entire Earth, but there is no widely accepted explanation for how or when it might have broken up.

A 2012 study of isotopes in the oldest rocks and minerals shows a considerable shift in their make-up around three billion years ago, which might indicate the point at which tectonics started. In 2015, the Russian-Swiss geophysicist Taras Gerya published modelling results which showed that at this time the mantle was 100–300 °C hotter than it is now. This would have resulted in weaker, more easily broken tectonic plates, so perhaps there was a larger number of small plates three billion years ago. But how could such an arrangement have led to a sustained pattern of subduction? With subduction, the leading edge of oceanic crust dips under the continental plate. For this to happen, the slab being pulled down must retain its integrity so that it drags the rest of the plate along with it. If it snaps off too easily, the ocean floor won't move continuously towards the coast. Without the 'pull' on the ocean floor, the system would break down. The puzzle about how subduction started and continued to keep going remains a challenge to geological theory and modelling.

Drifting continents

When we understand that plate movement is driven by subduction, it explains Wegener's theory of continental drift and,

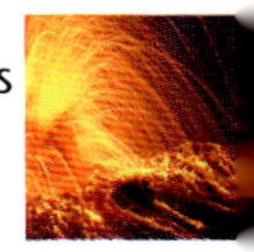

furthermore, suggests a pattern. Over time, the expanding ocean forces all the land-bearing plates together into a supercontinent. Then that supercontinent is broken apart by another

The Isua Greenstone Belt in southwestern Greenland is made up of some of the oldest known rocks in the world. Its tectonic history has been dated to 3.7–3.6 billion years ago.

EARTH'S FRACTURED SURFACE

Currently, Earth's crust is divided into seven major plates and dozens of smaller ones. The largest, the Pacific Plate, is estimated to be 103,300,000 sq km (39,884,353 sq miles) in size. The oceanic crust is about 7–10 km (4–6 miles) thick and the continental crust is up to 70 km (43 miles) thick in mountainous regions. Oceanic crust is made mostly of basalt, which leaks out at the mid-ocean ridges; the less dense continental crust contains a lot of granite and andesite.

Plates meet at faults, which are often the site of dramatic geological activity. Constructive and destructive faults are associated with volcanoes; destructive faults and collision zones are also connected with mountain-building. Transform faults are associated with earthquakes.

Granite forms many of the most spectacular geological landscapes, including the cliffs of Yosemite National Park in the USA.

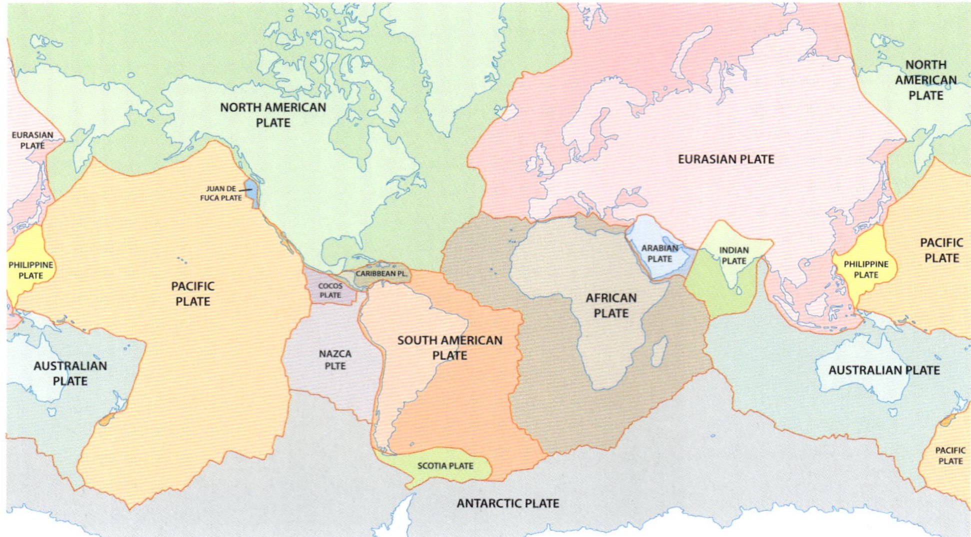

The larger tectonic plates and how they relate to landmasses and oceans.

spreading rift zone and separate continents appear again – for a while – until they merge once more.

Although the edges of landmasses are dynamic, accreting new materials and losing some rock, further inland the rock is stable and largely unchanging. The current continents are made up of arrangements of cratons with their accreted surroundings. The result is a patchwork, with cratons dotted around and their skirts of accreted rock joining them together (see page 59).

Drama and devastation

The shift of continental landmasses is very slow, and certainly not noticeable over a human lifespan. But tectonic movement also produces far more dramatic and often catastrophic natural events in the shape of earthquakes and volcanic eruptions, which the catastrophists perceived to be the opposite of Hutton's slow processes. We now know, of course, that they are all the result of slow processes.

Devastating geological events can easily kill large numbers of people and destroy towns or even entire civilizations. They naturally inspire awe and fear and, for millennia, people struggled to explain them. Inevitably, the earliest accounts placed them firmly in a mythological context; many people believed it was the way in which angry or vindictive gods rained down their judgement on humanity.

Collecting earthquake data

In Ancient Greece, Aristotle attempted a scientific explanation of earthquakes, describing winds within Earth which cause the surface to tremble. He gathered data about earthquakes and based his theory upon it. It is not only a fine example of his

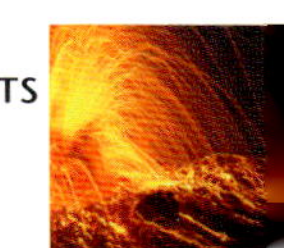

The Temple of Poseidon at Cape Sounion in Attica, Greece. One of the 12 Olympian gods in Ancient Greek myth, Poseidon was god of the sea and of earthquakes.

A SEISMIC CATFISH

In 18th-century Japan, it was believed that earthquakes were caused by a giant catfish, Namazu, wriggling in the mud beneath the Japanese islands. In the fable, the god Kashima steadies the catfish, pushing it against the foundations of Earth with a heavy stone to keep it still. But if Kashima is inattentive, the fish can wriggle and wreak havoc. In the 19th century, Namazu's behaviour was considered a punishment for human greed. The catastrophes that followed Namazu's activities led to a redistribution of wealth in Japan.

proto-scientific method, but the earliest recorded use of statistical methods in this way – but it was wrong.

Aristotle began by refuting three previous hypotheses, giving a useful summary of contemporary thinking. Anaxagoras, he said, argues that there are pockets of *aether* (the refined matter the Greeks thought made up the heavens) trapped within the Earth; these pockets escape, causing quakes on their way up and out. Aristotle dismissed this as 'too primitive to require refutation' because it assumes there is an 'up' on a sphere, and because it doesn't explain why only some regions have earthquakes. Aristotle also dismissed the account of Democritus: that earthquakes happen when rain falling on an already saturated Earth 'forces its way in', or when water rushes suddenly from wetter to drier areas below ground. Finally, Aristotle refutes the explanation of Anaximenes: that earthquakes happen as drying ground breaks up and compacts, or when heavy rain destroys the cohesion of the ground. Why then, Aristotle asks, does the occurrence of earthquakes not match a region's propensity for droughts or floods? This urge to match the explanation to the observed details was an important step in seeking a valid scientific explanation.

Aristotle believed evaporation to be the source of earthquakes. He observed that rain soaks into the ground, then the heat of the Sun and Earth causes it to evaporate and escape. Aristotle maintained that this produced wind, as air replaces the evaporating water. To support his theory, he cited as evidence the frequency of earthquakes in places where the ground

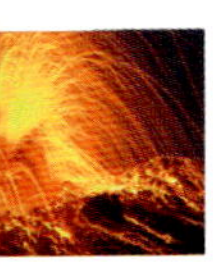

is 'spongy', such as Sicily. While there undoubtedly is spongy volcanic rock in Sicily, we now know that tectonic activity produces the earthquakes (and the rock). Aristotle noted a link between earthquakes and tsunami (see box), but also made spurious connections between when earthquakes occur and the time of day, cloudiness of the skies, and other aspects of weather.

By the time Aristotle came up with his theory, Chinese chroniclers had been noting earthquakes for nearly 2,000 years. Consequently, China has the longest historical record of earthquake activity. Most accounts, noted by local scribes, are fleeting. The earliest, from 2300 BC, notes 'earthquake and springs gushing'. In AD 977, Chinese earthquake data was gathered together in a text listing 45 earthquakes between 1100 BC and AD 618.

The single most significant development in earthquake monitoring came in the 2nd century AD, when the polymath Zhang Heng (AD 79–139) invented the first seismoscope – an instrument for detecting distant earthquakes. Although his seismoscope has not survived, and there are no detailed plans of it, his account has led to several reconstructions.

> *'The combination of a tidal wave with an earthquake is due to the presence of contrary winds. It occurs when the wind which is shaking the earth does not entirely succeed in driving off the sea which another wind is bringing on, but pushes it back and heaps it up in a great mass in one place. Given this situation it follows that when this wind gives way the whole body of the sea, driven on by the other wind, will burst out and overwhelm the land.'*
>
> Aristotle, *Meteorology*

TOWARDS AN EXPLANATION

Collecting data revealed that some areas are more prone to earthquakes than others, but it did not reveal the causes of earthquakes or enable people to predict them. Indeed, we still can't predict earthquakes well.

The first step towards a modern scientific understanding came in 1910 when American seismologist Harry Fielding Reid proposed an 'elastic rebound theory' following his careful studies of a catastrophic earthquake in San Francisco in 1906. He suggested that earthquakes occur when energy has built up along a

> *'The chief cause of earthquake is air, an element naturally swift and shifting from place to place. As long as it is not stirred, but lurks in a vacant space, it reposes innocently, giving no trouble to objects around it. But any cause coming upon it from without rouses it, or compresses it, and drives it into a narrow space . . . and when opportunity of escape is cut off, then "With deep murmur of the Mountain it roars around the barriers", which after long battering it dislodges and tosses on high, growing more fierce the stronger the obstacle with which it has contended.'*
>
> Zhang Heng, AD 132

Zhang Heng's ingenious and ornate seismoscope indicated the direction of an earthquake. Eight dragon-heads are arranged around the circumference of a bronze vessel, each one holding a ball in its jaws. The heads are connected to a crank and right-angle lever. When an earthquake shakes the ground, the mechanism works the crank, causing one of the dragons to drop its ball into the mouth of a bronze toad beneath, showing the direction of the seismic shock.

fault line and is suddenly released. This still forms the basis of our understanding of earthquakes at transform faults such as the San Andreas Fault in California.

Fire inside and out

Volcanoes are the other major tectonic cataclysm, and readily lend themselves to explanations involving gods and a fiery hell located within Earth. The Roman poet Virgil wrote that the giant Enceladus was buried beneath

In 1906, a devastating earthquake in San Francisco, intensity 7.9, killed 3,000 people. The photo shows a fissure in East Street after the earthquake. A carriage fell into cracks in the roadway near the waterfront caused by lateral spreading in the area.

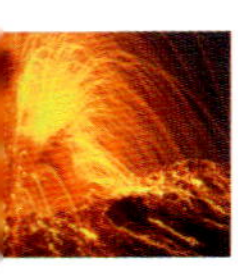

At the San Andreas Fault, the North American plate grinds against the Pacific plate along a 1,200–1,300 km (750–810 mile) tectonic boundary.

THE WORST EARTHQUAKES

Although we can now measure the intensity of earthquakes and compare them quantitatively, it's difficult to reconstruct those of the distant past. One measure of the severity of a historical earthquake is the number of casualties. Obviously, this is not a very scientific measure: few people will die if an earthquake strikes in an unpopulated area, and many will die if it strikes a city.

RANKING	WHERE	WHEN	DEATHS	MAGNITUDE
1	Shensi, China	23 Jan 1556	830,000	c.8
2	Tangshan, China	27 July 1976	255,000 (official) to 655,000 (unofficial)	7.5
3	Aleppo, Syria	9 Aug 1138	230,000	?
4	Sumatra, Indonesia	26 Dec 2004	227,898	9.1
5	Haiti	2 Jan 2010	222,570	7

Mount Etna in Sicily as a punishment for rebelling against the gods. When the mountain rumbled, Enceladus was said to be crying aloud in torment; the flames from the mouth of the volcano were his breath and the earth around the mountain juddering was when he shook the bars of his prison. Another giant, Mimas, was said to be buried under Mount Vesuvius near Naples in southern Italy.

Volcanic eruptions often come as a devastating surprise. Many volcanoes lie inactive (dormant) for hundreds or even thousands of years, so the eruption seems to come from nowhere. Volcanic soil is typically very fertile, so settlements often grow up nearby. When an eruption of Mount Vesuvius destroyed the towns of Pompeii and Herculaneum in AD 79, the scholar Pliny the Younger wrote: 'Many [people] besought the aid of the gods, but still more imagined there were no gods left, and that the universe was plunged into eternal darkness for evermore.'

This 19th-century depiction of the eruption of Vesuvius shows how the eruption of AD 79 would have looked. Volcanoes are of different types, but each volcano has similar eruptions, even over extended periods.

Houses of fire

In 1638, Athanasius Kircher ventured into the crater of Vesuvius to investigate and take measurements of the temperature – a perilous undertaking as the volcano was in an eruptive phase and still hot. Kircher formulated an explanation for volcanoes which became part of his theory about the inside of the Earth.

Kircher supposed that Earth's interior contained a linked series of fire-houses or 'pyrophylacia'. The largest and most important of the fire-houses was at the centre of the Earth and was the location of Hell – a startling combination of geology and religion. He believed Purgatory

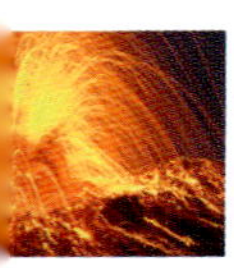

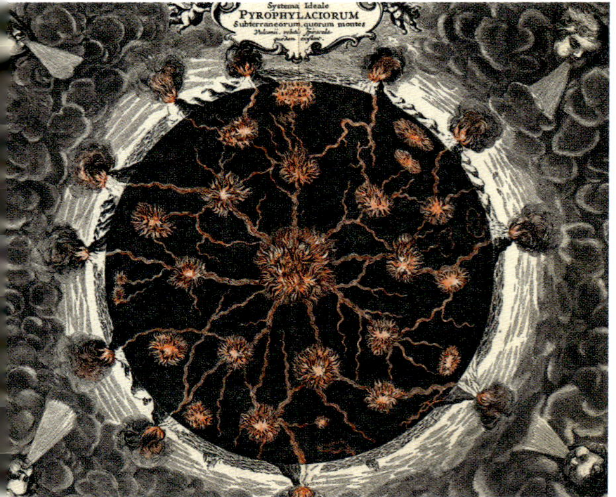

'In the middle of the night I climbed the mountain by hard and rugged paths. When I reached the crater, horrible to relate, I saw it all lit up by fire, with an intolerable exhalation of sulfur and burning bitumen. Thunderstruck by the unheard-of spectacle, I believed I was peering into the realm of the dead, and seeing the horrid phantasms of demons, no less, perceived the groaning and shaking of the dreadful mountain, the inexplicable stench, the dark smoke mixed with globes of fire which the bottom and sides of the mountain continuously vomited forth from eleven different places, forcing me at times to vomit out myself. . . . When dawn broke, I decided to explore diligently the whole of the interior constitution of the mountain. I chose a safe place where a firm foothold might be had, and descended to a vast rock with a flat surface to which the mountain slope gave access. There I set up my Pantometer and measured the dimensions of the mountain.'

Athanasius Kircher, 1664

Top left: *Kircher's conception of Earth's interior had linked fire chambers.*

Bottom left: *The Phlegraean Fields is a volcanically active area of Italy, near Naples. It has hot ground and fumeroles that reek of sulphur and emit hot gases constantly. The Roman poet Virgil wrote that the blood of giants defeated in their war with the gods rose to the surface in the Phlegraean Fields.*

MYTHICAL DEVASTATION

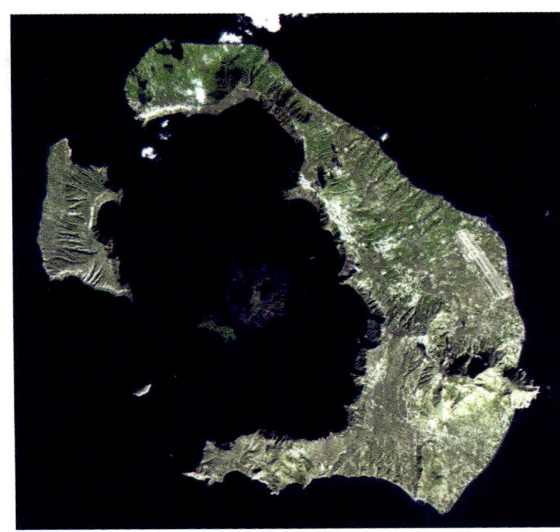

The shape of Santorini reveals its origins as a devastated volcano.

The Minoan eruption, which occurred between 1642 BC and 1540 BC, destroyed part of the island of Akrotiri (Santorini) and possibly ended Minoan civilization. It was one of the largest eruptions in human history, with an intensity of 6 or 7 and triggering a tsunami up to 150 m (492 ft) high. The volcano had already built up and destroyed itself several times over a period of many hundreds of thousands of years. Today, the island of Santorini has a shape characteristic of a sea-filled volcanic caldera (crater) with a ring of rock. The island's destruction has been linked (inconclusively) with Plato's account of the lost city of Atlantis. There are accounts of events that relate to the eruption from as far away as China, where 'yellow fog, a dim sun, then three suns, frost in July, famine, and the withering of all five cereals' accompanied the fall of the Xia dynasty in 1618 BC.

to be located at an intermediate point and remarked that monks in a monastery near the Phlegraean Fields, a large supervolcano west of Naples, could hear the groans of sinners from beneath the ground.

Kircher supposed that on the third day of Creation, when God separated the land and sea, He created chambers within Earth, called 'geophylacia'. These were of three types: fire-houses, air-houses and water-houses. He thought another kind of chamber contained the 'seminal principles' which enabled the growth of minerals underground.

Kircher believed that water-houses lay beneath mountains and provided the water for springs and rivers. As this source was finite, he supposed Earth's water was sucked back into the water-houses at whirlpools, such as the Norwegian maelstrom, and fed back into the rivers and streams. He believed the largest and most important whirlpools resided at the poles. According to his model, the water-houses were all linked and the pressure of the tides acted like a bellows, forcing water to flow through the channels within Earth and emerge in springs, replenishing lakes and seas. As

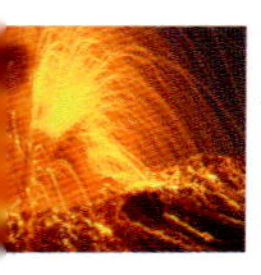

water flowed inside the Earth, it nourished the growth of crystals and minerals. Kircher thought that mountain ranges formed a structural skeleton for Earth.

Other ideas from the 16th and 17th centuries included volcanoes which expelled Earth's waste material as tears and excrement in the form of bitumen, tar and sulphur (Johannes Kepler); volcanoes that formed when the Sun's rays pierced the three-layer structure of the Earth – comprising air over water over the fiery depths (René Descartes); and vapour which, under pressure, produced eruptions (Agricola).

Kircher thought that water re-entered Earth through the Saltstraumen Maelstrom, off the coast of Norway.

Building volcanoes

The Scottish geologist Charles Lyell was the first to propose that volcanoes build up slowly; he called this 'backed-up building'. The conventional view was that they emerged quickly as the result of rapid upheaval. In fact, some volcanoes emerge quickly and others emerge slowly.

There are two main types of large volcano:

Stratovolcanoes or **composite volcanoes** are steep-sided volcanoes which produce thick lava and pyroclastic flows (a fast-travelling, superheated mix of ash, rock, dust and steam). The cone is built up of layers of ash and hardened lava from previous eruptions. Stratovolcanoes occur near subduction zones and erupt infrequently but violently. Sometimes, small fast-growing volcanoes called cinder cones grow on the flanks of stratovolcanoes. Famous stratovolcanoes include Mount St Helens and Vesuvius.

Shield volcanoes are low and shallow with gently sloping sides. They erupt frequently but not explosively, producing runny lava that flows far over the ground before cooling. They occur at hotspots and constructive boundaries; the volcanoes of Hawaii are shield volcanoes.

A few volcanoes have had eruptions rated at 8 (the highest) on the Volcanic Explosivity Index (VEI). This describes eruptions that release at least 1,000 cubic km (240 cubic miles) of deposits. These supervolcanoes generally have no mound (more often a depression) and are hard to identify. They seldom erupt – the most recent was the Oruanui eruption of New Zealand's Taupo volcano, which occurred

26,500 years ago. The last major eruption of the supervolcano under Yellowstone Park in the USA was 640,000 years ago.

Forged by fire

Volcanic eruptions can have an immense impact on the environment, affecting rock, atmosphere and the living world. They have been implicated in some of the most devastating mass extinction events in Earth's history (see pages 172–4), but they also mould the landscape and the climate.

A VOLCANO FROM NOWHERE

The cinder cone Paricutín emerged from a farmer's cornfield in Mexico in 1943. For a few days beforehand, local people heard low rumblings like thunder, indicative of deep earthquakes. Three hundred of these low-intensity earthquakes took place on the day before the volcano began to appear. A local farmer, Dionisio Pulido, watched as a mound rose from a crack in his field. It first appeared at 4 pm and by nightfall was shooting flames 800 m (2,625 ft) into the air. By the end of a week, a cone of rock and ash 100–150 m (328–492 ft) tall had accumulated in Pulido's field. Ash, lava flows and semi-molten lumps of rock rained down on the area, forcing local people to evacuate the towns of Paricutín and San Juan Parangaricutiro. These towns were eventually buried in lava, and remain so. The cinder cone grew to 365 m (1,197 ft) high in eight months.

The church of San Juan Parangaricutiro near Paracutín is stranded amidst a solidified sea of lava.

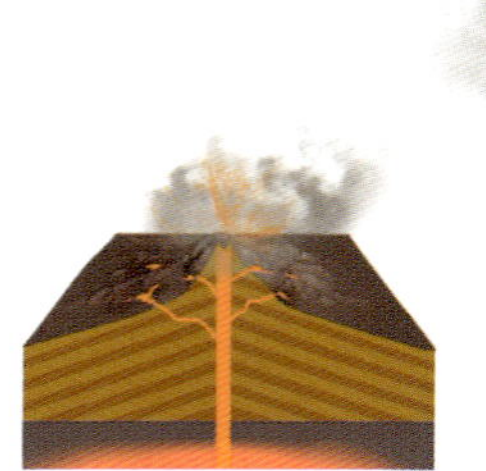

From left: lava escaping between separating tectonic plates through a fissure vent; fast-moving lava flows forming the layers of a shield volcano; slower-moving lava forming a steep-sided lumpy 'dome' with vents (a lava dome); and a cone shape with explosive eruptions and lava interleaved with volcanic ash (a stratovolcano). The magma chamber of each volcano is partially seen across the bottom.

In 2017, geologists in Canada and Russia produced a database of Earth's most cataclysmic volcanic eruptions. They identified some dating back more than two billion years and they plotted the extent of the lava flows from each.

These eruptions were world-changing. One, in Siberia, 252 million years ago, is linked with the most severe extinction event in which 90 per cent of species died out (see page 172). Although the lava floods left by these volcanoes have mostly been eroded away, dyke swarms remain as evidence. These are the channels through which lava has spread, fanning out from the main throat of the volcano. The age of the eruptions was established using uranium–lead radioactive dating of these remnants.

Huge eruptions can continue for millions of years, pouring out over a million cubic kilometres of lava during that time. They seem to happen about every 20 million years (the last was 17 million years ago).

Under the volcano

Massive eruptions flood the land with basalt lava that can be kilometres thick. Examples of such events include the Siberian Traps in northern Russia, 252 million years ago and the Deccan Traps in India, 66 million years ago. These eruptions alter the geology of the land not just by covering it with a new layer of rock but by changing the rock that is already there. Boulders are picked up and carried by lava flows, but even more significantly existing rock is baked by the heat of the eruption or flowing lava. Sedimentary rock undergoes a chemical change into metamorphic rock; for example, when volcanic heat bakes limestone, it changes it to marble. Volcanic heat can also burn peat and coal out of the ground.

Blowing hot and cold

Computer models from 2017 suggest that, at their height, the violent Siberian eruptions could have raised global temperatures by 7°C with their emission of hot gases. But the temperature can soon plunge, as sunlight is blocked by ash and dust high in the atmosphere. This process was seen most recently in the 'year without a summer' of 1816, which followed the eruption of Tambora in Indonesia (see box on page

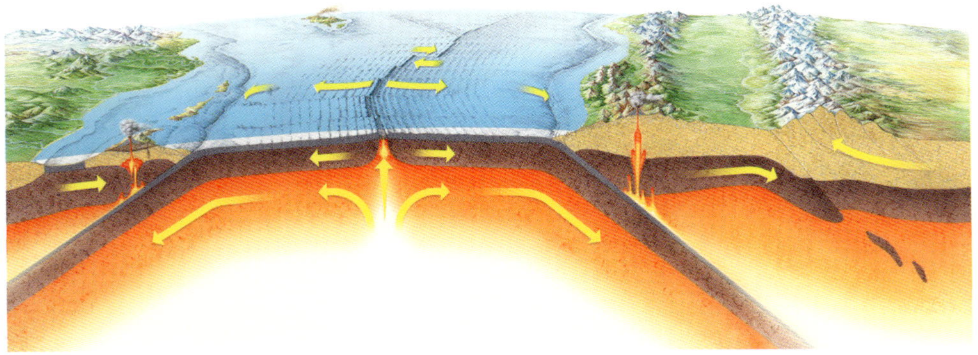

Tectonic activity around an ocean basin includes the emergence of new rock at the mid-ocean ridge and the subduction of old rock near the coast. The subduction zone on the left is off-shore, while the one on the right is coastal. Continental crust (light brown) and oceanic crust (blue) lies on top of heavier (purple) rock that slowly moves, dragging the crust with it. Subducted rock melts and feeds volcanoes along the subduction zone. Subduction pulls the sea floor towards the coast. This, with the emergence of new rock mid-ocean, drives sea floor spreading and continental drift.

112). Ash and sulphur dioxide are washed out of the atmosphere quite quickly by rain. But rain containing dissolved sulphur is acidic and has its own impacts. The longer-term effect of extensive eruptions is climate warming. As sedimentary rock bakes, organic material in the rock burns, releasing methane. Even baking carbonate rocks with no organic material can release carbon dioxide. Massive eruptions also pour out carbon dioxide with magma along with halogens that destroy ozone and let harmful radiation from the Sun penetrate the atmosphere. The acid rain that follows an eruption dissolves carbonate rocks, releasing yet more carbon dioxide.

Making mountains

Not every mountain is volcanic, built by magma from below. Many are built from rock that is already solid. This happens at destructive faults where two continental plates collide. The vast mountain ranges of the Himalayas, the Alps and the Andes have formed in this way.

MECHANICS OF MOUNTAIN-BUILDING

A collision between continents begins when they are separated by ocean, and when the

> *'Had the fierce ashes of some fiery peak*
> *Been hurl'd so high they ranged about*
> *the globe?*
> *For day by day, thro' many a blood-red*
> *eve . . .*
> *The wrathful sunset glared'*
>
> Alfred, Lord Tennyson, 'St. Telemachus', published in 1892 in the aftermath of brilliant sunsets caused by the eruption of Krakatau in 1883.

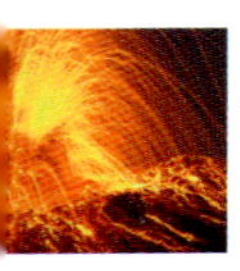

ocean is no longer widening. The oceanic crust is subducted beneath one of the plates, raising volcanoes some distance beyond the subduction zone. But when all the oceanic crust has been subducted and just the lighter crust of the continent is left, subduction becomes much more difficult. The continental crust is lighter than the mantle and subducts into it more slowly, if at all. It is no longer feeding volcanoes, which dry up and the crust buckles and piles up at the collision zone. The top plate is pushed upwards, folding and deforming as the plates move towards each other.

Rock at the edges of the converging plates is at first compressed. The immense

THE YEAR WITHOUT A SUMMER

The largest eruption in recorded history was in 1815. The volcano Tambora on Sumbawa, Indonesia, poured so much ash into the air that it blocked out the Sun. The ash rose to the stratosphere and was carried around the world, causing disastrous cooling. Snow fell in June in New York; outside Quebec, Canada, the snow lay 30 cm (12 in) deep. Crops died from lack of sunlight, farm animals starved, and famine and disease spread through the vulnerable population. In China and India, disruption to the monsoon system caused catastrophic flooding. The following winter was extremely harsh and affected the year's harvest.

The eruption might even have prompted the invention of the bicycle. The horses that were vital to transportation at the time were fed a diet of oats. The global shortage of this cereal could have inspired the German inventor Baron Karl von Drais to investigate a method of horseless transport in 1817.

The eruption of Tambora in 1815 devastated the island of Sumbawa.

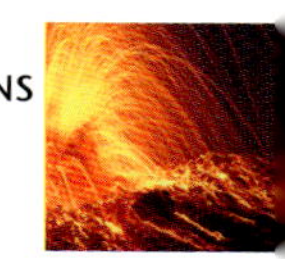

pressure exerted on the rocks changes them, producing metamorphic rock. Where the heat of subduction and pressure melts rock, the molten rock does not rise through volcanoes but freezes in the rock above and forms blobs of intrusive igneous rock, called plutons. Metamorphism also happens locally around plutons where their heat bakes adjoining rock. All together, these processes produce a rich mix of rock and structures which tell the story of how the mountains have formed.

At the same time, rocks near the surface ruck up, like a cloth pushed across a table, forming wrinkles and folds. When further pressure can't be absorbed by folding, then faults occur: the rock breaks and slabs are thrust out of the way. At a fold, layers are continuous but curved. At a fault, the layers are discontinuous as a whole slab breaks apart and shifts. The result is the often odd arrangement of layers of rock seen in mountains. Layers sometimes end up vertical, and frequently show extravagantly convoluted folds and clear faults.

The crust thickens where plates are forced together, not only pushing up the mountain but also thickening the lower edge of the crust, giving mountains deep 'roots'. Mountains are heavy – they weigh down the crust, creating a dip at the outside edge of the range. Over time, this fills with sediment produced by the weathering and erosion of the mountains, and forms a sedimentary basin. Alternatively, sediment is carried away by glaciers, streams and rivers to alluvial plains and deltas.

The mountains grow slowly and for as long as the plates are forced together. When the direction of the plates' movement changes, the mountains are left as a scar running down the middle of a continental landmass.

Torn apart

Where land pulls apart at a continental rift, volcanic mountains emerge at points of thin crust where magma can well up. Even non-volcanic mountains can form at rifts. The forces at work pulling the plates apart can break great slabs of crust, causing a portion of the surface to drop, giving a depression or 'rift'.

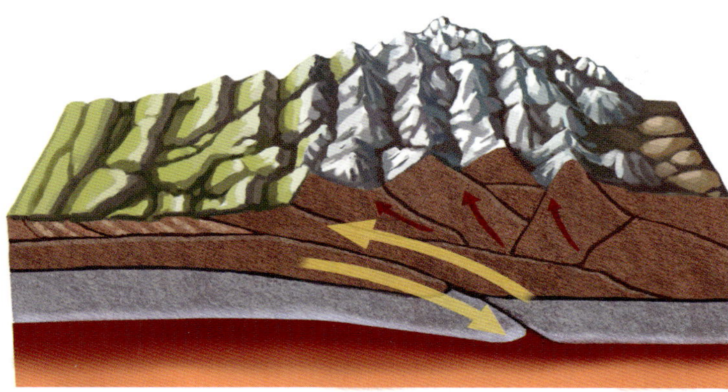

By riding over the Indian plate, the Eurasian plate (right) formed the Himalayas.

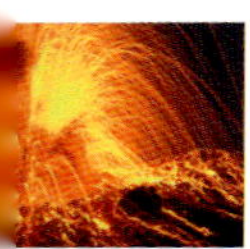

AS OLD AS THE HILLS

Some mountain ranges are very old. The Appalachians began to grow when the North American and African plates collided during the formation of Pangea 450 million years ago, making the Central Pangean Mountains. They finished building 250 million years ago and are now eroding. At their highest, they were taller than the Himalayas. The Scottish Highlands and the Little Atlas Mountains of Morocco are also relics of this range, split apart by the widening Atlantic.

On the way down

Once mountain building is finished, the mountains begin to be worn away by the effects of weather and erosion, but it is not a straightforward matter of losing height. A mountain is a heavy lump of rock which deforms the lithosphere so that it dips down into the mantle below the mountain's base. As bulk is removed, the mountain becomes lighter and there is a certain amount of rebound as the lithosphere rises back up. Although it may be less pointed where it has been smoothed down by weathering, the mountain might be just as tall, or nearly as tall, as before the rebound of the lithosphere added to its height. The removal of mountains is a

slow process. The Makhonjwa Mountains on the border of South Africa and the Kingdom of Eswatini (previously Swaziland) are 3.5 billion years old, yet the highest is still 1,800 m (5,900 ft) tall.

The brilliant colours of the Rainbow Mountains of Zhangye Danxia National Park, China, are produced by oxides in the sandstone. The layers of rock originally laid down horizontally were tilted at an angle by faulting as the mountains formed.

Facing page: *The East African rift valley in Ethiopia. The broken slabs of crust that remain create block mountains.*

WATCHING MOUNTAINS GROW

With the exception of a few fast-forming volcanoes such as Paracutín, mountains grow too slowly for us to notice them, even over centuries. Yet GPS systems accurate to within a couple of millimetres (about a 12th of an inch) now allow geologists to measure the horizontal narrowing and vertical growth of mountains as they build. Measurements show that the Andes are narrowing by 2 cm (0.8 in) a year and growing vertically by 2 mm (0.08 in). This means they have grown a little over 1 metre (3 ft) since the demise of the Aztec civilization 500 years ago.

CHAPTER 6

Life changes EVERYTHING

'Organic life beneath the shoreless waves
Was born and nurs'd in ocean's pearly caves;
First forms minute, unseen by spheric glass,
Move on the mud, or pierce the watery mass;
These, as successive generations bloom,
New powers acquire and larger limbs assume;
Whence countless groups of vegetation spring,
And breathing realms of fin and feet and wing.'

Erasmus Darwin, *The Temple of Nature*, 1802

Unlike other planets in the solar system, Earth is home to abundant and diverse forms of life. While we can't be certain that all of our planetary neighbours are uninhabited, we know that life on Earth has helped to shape the planet we live on.

Anemones at the Beebe hydrothermal vent field in the Caribbean. Life on Earth might have begun in places like this.

First life

Life on Earth began with organic (carbon-containing) chemicals which could, in the right conditions, make copies of themselves.

The carbon-based molecules essential to building living things are often referred to as prebiotic. They include amino acids, which are the building blocks of proteins. In the right conditions, prebiotic molecules can be built from elements found in abundance on Earth and elsewhere in the universe: carbon, hydrogen, nitrogen and oxygen. Prebiotic molecules could have been created on Earth in the hot conditions of thermal pools, as the result of lightning strikes, or in undersea volcanic vents. Or they could have been delivered to Earth on meteorites from Mars or elsewhere – even from outside our solar system. Or perhaps there was a mix of home-grown and imported prebiotics.

If prebiotic molecules can be carried through space on meteors, they could have seeded many environments besides Earth. Similarly, if they arose with little difficulty in the right conditions on Earth, they could perhaps arise just as easily elsewhere.

> *'So with animals, some spring from parent animals according to their kind, whilst others grow spontaneously and not from kindred stock; and of these instances of spontaneous generation some come from putrefying earth or vegetable matter, as is the case with a number of insects, while others are spontaneously generated in the inside of animals out of the secretions of their several organs.'*
>
> Aristotle, *On the History of Animals*, Book V, Part 1

Hydrothermal vents such as this in the Atlantic Ocean produce a warm undersea environment enriched with minerals. It is an environment in which simple life can thrive.

Life from nowhere

For many centuries, people believed that some forms of life could emerge from inanimate matter – a model now known as spontaneous generation. It was used to explain how food left to rot becomes infested with maggots or how mice can turn up in containers of corn.

The theory of spontaneous generation was challenged in 1668 when the Italian physician Francesco Redi showed that maggots appear in meat only if flies have access to it. Even so, in 1809, the French biologist Jean-Baptiste de Lamarck was still proposing that: 'Nature, by means of heat, light, electricity and moisture, forms direct or spontaneous generation at that extremity of each kingdom of living bodies, where the simplest of these bodies are found.'

In 1871, Darwin wondered more specifically 'if (& oh what a big if) we could conceive in some warm little pond with all sort of ammonia & phosphoric salts, – light, heat, electricity &c present, that a protein compound was chemically formed, ready to undergo still more complex changes.'

Under the microscope

In 1922, the Russian biochemist Alexander Oparin began with the fundamental proposal that there is no material difference between living and non-living matter, and that life depends on the chemical behaviour of molecules. The discovery of methane in the atmosphere of Jupiter led him to suggest that early Earth had a strongly reducing atmosphere (one in which oxidation does not take place). He thought the early atmosphere probably contained methane, ammonia, hydrogen and water vapour and that these may have provided the building blocks for life. Oparin suggested a process

A vesicle, with an isolated internal environment.

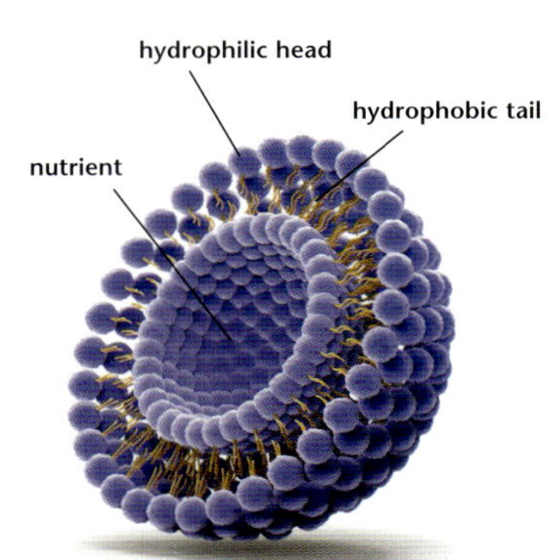

CAPTURING AN INSIDE

Coacervates are microscopic spontaneously formed droplets which can isolate an internal environment. They include micelles, balls formed from molecules which have a hydrophobic (water-shunnning) part and a hydrophilic (water-loving) part. In water, the hydrophobic parts cluster together with the hydrophilic parts on the outside of the ball. Micelles can combine to form vesicles – globules which have two layers of molecules forming a wall around a cavity. The hydrophobic ends of the molecules are in the inside of the wall and the hydrophilic ends form the internal and external surfaces. A vesicle can isolate an internal environment in the central cavity, which can become different from the external environment.

that started with the simplest of organic molecules and formed progressively more complex systems, perhaps developing from coacervates (see box on page 119).

Oparin suggested that abiogenesis (life from inanimate matter) came about through a serendipitous mix of chemicals. His findings were corroborated in 1929 by the American biologist John Haldane, who said that the constituents 'accumulated till the primitive oceans reached the consistency of hot dilute soup'. The early, life-spawning chemical mix is commonly referred to as the 'primordial soup'.

Making soup

In 1953, a University of Chicago graduate student, Stanley Miller, and his professor, Harold Urey, set out to recreate conditions on early Earth.

Starting with Oparin's proposal of an atmosphere rich in water, methane, ammonia and hydrogen, they combined this cocktail in a sealed system. They then heated the 'ocean' they had made to replicate the early atmosphere and exposed the gas and steam to a stream of electric sparks to simulate the lightning strikes thought to be common on early Earth. They cooled and condensed the atmosphere, allowing it to 'rain' back into the ocean. After just a week, 10–15 per cent of the carbon had produced organic compounds,

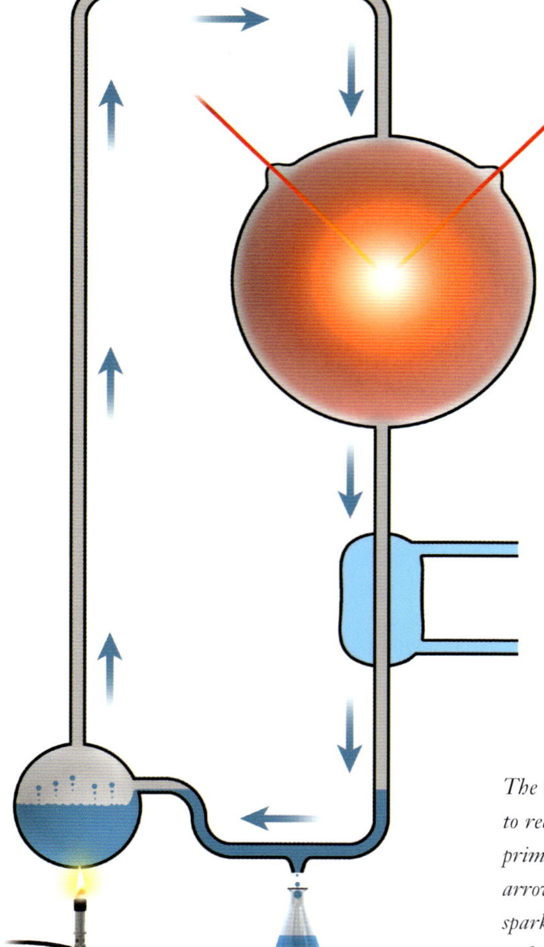

The equipment used for the Miller-Urey experiment to recreate conditions on early Earth. A simulated primordial ocean is heated in the flask. The blue arrow indicates the path of gas and steam. Electric sparks are applied to this 'atmosphere', which is then cooled in a condenser. Samples are collected; the rest of the condensate returns to the 'ocean'.

and 2 per cent was in the form of amino acids. The 'ocean' also contained purines and pyrimidines needed to make RNA and DNA, the chemicals which carry the genetic code of life.

In 1961, Juan Oro found he could produce amino acids, including large quantities of adenine, from hydrogen cyanide and ammonia dissolved in water. Adenine is one of the four bases of DNA and RNA and a key ingredient of ATP (adenosine triphosphate), a chemical crucial to storing and releasing energy in cells.

Making the jump

To make the transition from organic chemicals to life, matter had to develop strategies for self-organization and replication. It's now thought that the first life forms used RNA rather than DNA to hold their genetic code. RNA is simpler than DNA, as it is a single strand of nucleotide bases rather than two strands joined into a double helix. Life on Earth might have come from a hypothesized 'RNA world' of self-replicating molecules.

One hypothesis suggests that prebiotic molecules or even life on Earth itself originated in space. Prebiotic molecules, including amino acids, can be carried on asteroids and meteors. Some organic molecules might have been present in the dust of the protoplanetary disk and could have been incorporated into planets from their very beginnings. This idea that life originated in space is called 'panspermia'. In 1783, the French natural historian Benoît de Maillet wrote of life starting from 'germs' (as in seeds, not pathogens) falling into the oceans from space.

'We must regard it as probable in the highest degree that there are countless seed-bearing meteoric stones moving about through space. If at the present instance no life existed upon this Earth, one such stone falling upon it might, by what we blindly call natural causes, lead to its becoming covered with vegetation.'

Lord Kelvin, 1871

There are three variants of panspermia theory. Life can come from within the solar system (interplanetary or 'ballistic' panspermia), from outside the solar system (interstellar or 'lithopanspermia') or be deliberately seeded by intelligent beings in space (directed panspermia). In the first two, it arrives incidentally on asteroids or other bodies, but in the last it is spread deliberately.

Support for the extraterrestrial origins of life on Earth was boosted by the arrival in Australia of the Murchison meteorite in September 1969. In 2010, analysis of a chunk of it found 14,000 different molecular compounds, including 70 amino acids; researchers think it might in fact contain millions of organic compounds. The interior of the meteorite is pristine, leaving no doubt that prebiotic molecules can survive in space and be delivered to Earth, but it gives no clue as to whether this was an important source of prebiotics on early Earth.

Right from the start

The fossil record requires rocks, so obviously we can't find any fossils older

The Murchison meteorite weighed over 100 kg (220 lbs), representing a large chunk of primordial rock from space.

Morning Glory mineral-rich hot spring in Yellowstone National Park, where archaea flourish.

than the oldest rocks available. The impact which created the Moon 4.52 billion years ago would have melted the entire surface of the planet, sterilizing it completely. This doesn't mean there was no life before the impact – just that it would have had to start again afterwards.

The oldest uncontested evidence of life on Earth dates from 3.5 billion years ago. However, chemical 'signatures' and some structures, including those in sedimentary deposits laid down in Greenland 3.8 billion years ago, indicate that life may be much older. The oldest trace of possible microbial life has been found on rocks in Canada which precipitated from hydrothermal vents between 4.28 and 3.77 billion years ago. The earliest organisms had distinctive chemicals in their cell membranes which are not readily broken down, so their presence is a good indicator of life.

ARCHAEA APPEAR

Until the mid-20th century, living organisms were generally categorized simply as plants or animals, groups that had long been recognized. When it became apparent that this didn't serve well, a five-kingdom system emerged consisting of animals, plants, bacteria, fungi and protists. It was something of a shock, then, when American microbiologist Carl Woese identified an entirely new type of organism in 1977.

While studying the DNA of bacteria, Woese discovered they fell into two clear and very different groups. One group, found at very high temperatures or producing methane, was genetically distant from other

bacteria and domains of life. Woese soon realized that it was not a bacterium at all, but a specific type of primitive organism, now known as archaea. The five-domain system has now been reconfigured to include archaea, prokaryotes (which encompasses other bacteria) and eukaryotes (which encompass the other four kingdoms).

The first organisms on Earth might have been archaea. They flourish in extreme environments and could have been as happy in Darwin's 'warm pond' as in a scorching deep-sea vent. But there could have been something which predates archaea and possibly gave rise to prokaryotes and eukaryotes too. This earliest ancestor might not even have had a cell membrane.

Making changes

Whenever and wherever life began, it changed the immediate environment. As microbes died, they were deposited on the ocean or pool floor, creating the first organic sediments and leaving the first fossil evidence. Sedimentary rocks began to incorporate not just earlier rocks that had been ground up, carried and deposited, but the tiny remains of previously living things.

Around 3.5 billion years ago, stromatolites began to form. Stromatolites are made up of microbial mats – extensive colonies of microbes living in close proximity – together with the particles they have trapped, which are laid down in layers and form humped rocks. Stromatolites are still forming in some parts of the world, including Shark Bay in Australia, where they were first identified in 1956. They grow at the rate of about 1 cm (0.4 in) every 25 years.

Photosynthesis

Early methanogenic bacteria metabolized carbon dioxide and produced methane, giving Earth a methane-rich atmosphere. Then, between two and three billion years ago, microbes evolved that were able to photosynthesize. Like modern plants, they used energy from sunlight and carbon dioxide from the atmosphere, with water, to produce sugars and release oxygen. This set the stage for the modern atmosphere

Stromatolites revealed at low tide at Shark Bay in Australia. Similar fossil evidence from 3.2 billion years ago was found in South Africa in 2018.

which makes possible virtually all eukaryotic life, from amoeba to humans, blue whales and redwood trees. The exact date for the evolution of photosynthesis is still debated, with dates ranging over nearly a billion years, but it was the key to the beginnings of life as we know it on Earth.

Marine organisms that photosynthesize needed to live near the surface of the ocean, where sunlight could penetrate. They quickly became successful as there was plenty of sunlight and no competition for it. The methanogenic bacteria were forced downwards. They could still function at depth, since they didn't need the sunlight, but it meant the methane they released could not enter the atmosphere as readily.

The Great Oxygenation Event

The amount of oxygen in the oceans began to increase. At first it reacted with dissolved iron and was deposited as iron oxide (rust), which fell to the sea floor. As the sediment turned to rock, the iron oxide was incorporated into minerals such as hematite and goethite, leaving characteristic red bands in the rock.

Eventually, most of the iron in the oceans had been oxidized. Oxygen began to escape from the oceans into the atmosphere in a process known as the Great Oxygenation Event (GOE), dated to around 2.5–2.4 billion years ago.

Increased oxygen levels promoted algal bloom, involving the creation of even

All animals, even these blue whales, ultimately depend on the energy fixed through photosynthesis. Photosynthetic plants and microorganisms are at the bottom of every food chain.

more oxygen-producing organisms. As the organisms died and sank, the carbon-containing detritus of their cells was buried in the sediment, which meant there was less carbon available to be recycled as carbon dioxide. This reduction in greenhouse gases not only changed the composition of the atmosphere, but also cooled the planet.

The clear red bands indicate iron oxide in the form of hematite ore, which were created during the Great Oxygenation Event that began about 2.5 billion years ago.

The rise in oxygen levels had a catastrophic effect on life-forms flourishing in the oxygen-free environment. The oxygen was a poison to them, and the GOE triggered the first known mass extinction event – of anaerobic microbes.

Although the amount of oxygen increased dramatically, there was still far less than there is now. Measurements of different types of chromium oxides found in rocks from the mid-Proterozoic (1.8–0.8 billion years ago) shows only 0.1 per cent of current levels. It was enough to poison the previous organisms, but not enough for complex aerobic organisms such as animals to develop.

Greenhouses and snowballs

The oxygen-producing cyanobacteria soon had to contend with a mass extinction of their own – climate change on an unimaginable scale.

As the quantity of oxygen in the atmosphere increased, the amount of methane decreased, and oxygen reacted with the methane to produce carbon dioxide. An atmosphere rich in methane had helped to keep Earth warm (uncomfortably warm from our point of view). But as carbon dioxide is a much less powerful greenhouse gas, the warming effect was greatly reduced. It appears that the atmospheric change plunged the planet into the first of several 'Snowball Earth' phases, during which the temperature dropped so low that ice covered the entire surface.

The first evidence of extensive glaciation in the Paleoproterozoic era was discovered in 1907 in the rocks of Lake Huron in North America. Two layers of non-glacial deposits are sandwiched between three glacial layers laid down between 2.5 and 2.2 billion years ago. During the 20th century, evidence of glacial deposits during the same period emerged from as far afield as South Africa, India and Australia, and made it difficult to resist the idea that something widespread had occurred.

Dropstones

In the 1960s, the British geologist Walter Brian Harland proposed episodes of global glaciation after finding geological evidence of glacial action in rocks which would have been in the tropics 700 million years ago. Glaciers carry debris, from sand-sized grains to boulders, and when a glacier's underside melts, it leaves 'dropstones' in its wake. The first person to propose the idea of an ice age on Earth was the German philosopher-poet and statesman Johann Wolfgang von Goethe in 1784. Goethe suggested that the many large dropstones in the Alps had been transported by glaciers during a period of 'grim cold', when an ice sheet covered Germany. The idea that the entire planet had once been covered by ice

FREEZING COLD

The American geologist Joe Kirschvink coined the term 'Snowball Earth' in 1989. Earth has endured several periods of glaciation. Some have been global, 'Snowball Earth' events, others have been less severe, with ice covering only parts of Earth's surface, and are known as 'ice ages'. A glacial period needs at least one permanent ice sheet, which can be restricted to one or both of the poles. Most Snowball Earth events and ice ages have interglacial periods when the ice disappears and then returns.

The first ice ages occurred in the Archean eon. Then there were two ice ages either side of the first Snowball Earth event, which was associated with the Great Oxygenation Event. We are in a glacial period now and have been for the last 2.6 million years. Even though the climate is relatively warm, there are still ice sheets at the North and South poles. The last ice age is generally considered to have ended 11,000 years ago when the northern ice sheet retreated from most of Europe. In geological terms, however, we are still in the ice age, with a cooler phase of it having ended just recently. It seems likely that human activity will end the current ice age.

From space, Snowball Earth would have been almost entirely white, covered with snow and ice.

A dropstone of quartzite deposited by glacial movement in Itu, in the state of São Paulo, Brazil.

seemed so improbable that few people were willing to accept it. Calculations showed that the Sun would have needed to produce 1.5 times as much energy as it does today for Earth to escape its snowball state. And the Sun produced less energy in the distant past than it does today.

But it was tectonic activity, not the Sun's heat, that would bring an end to Earth's cold spell. Chunks of the seabed continue to be subducted at the edges of continents, regardless of the temperature, and volcanoes pour carbon dioxide out into the atmosphere. This is part of the slow carbon cycle. But another part of the cycle, the weathering of rocks which removes carbon dioxide from the atmosphere, slows and stops in very low temperatures.

Normally, acidic rain (water with dissolved carbon dioxide) falls on to silicate rocks and dissolves some of the surface, releasing calcium and bicarbonate ions. These come together in the ocean to produce carbonate rocks, which lock away carbon. When weathering is halted and volcanic activity continues, more carbon dioxide is released than removed. It builds up slowly in the atmosphere, starting a greenhouse effect which eventually tips into runaway warming. As ice melts, the planet's albedo (reflectiveness) reduces. This allows more heat from the Sun to be absorbed, so the planet warms ever faster.

Holed up

It might seem impossible that life could survive a Snowball Earth event; and it probably could not survive if all the water on the planet was frozen. Some scientists suspect that Earth might have been a slightly slushy snowball, perhaps with a band or pockets of liquid water near the equator.

Another theory is that ice did indeed cover the entire surface, but small areas of geothermal activity left cryoconite holes in it. These were filled with liquid water and enabled life to cling on until the end of the disaster.

Cryoconite holes form in areas of geothermal activity or when dust is deposited in ice or snow. The deposit is darker and therefore absorbs solar radiation better than the surrounding area.

New life for old

The Great Oxygenation Event resulted in the emergence of life-forms which could metabolize oxygen. These set off on new paths of development. One of the most significant changes was when single-celled organisms began to group together in colonies and became multicelled organisms. Cells differentiated to perform different functions and to rely on one another. Perhaps around 2–2.5 billion years ago, some cyanobacteria and possibly other similar organisms developed a simple form of multicellularity. The first definite multicellular organism, dated to 2.1 billion years ago, was found in Gabon, Africa, in 2008. The fossil is a flat disk with radial slits and a scalloped edge and measures 5 cm (2 in) across.

These strange scalloped disks represent some of the first multicelled organisms for which we have fossil evidence.

Evolutionary biologists originally believed that the leap from single to multiple cells must have been difficult to make, but recently it was found that it's quite easy to

prompt some single-celled organisms to become multicelled; the trick is to constrain their environmental conditions.

In an experiment in 2015, yeast cells were prompted to become multicellular colonies in just a week. They developed a 'snowflake' form, where new cells cling to the parent cell rather than breaking free. After 3,000 generations, a primitive form of multicellular reproduction starts, with branches being released as new colonies.

The 'Boring Billion' years

However easy or difficult it was to make the transition to multicellularity, it seems that a wide variety of more complex multicellular organisms didn't evolve for a long time. Paleontologists often refer to the years between about 1.8 billion and 800 million years ago as the 'Boring Billion'. It was a period of relative stability, both in terms of geology and evolution, which was essentially stalled with single cells in colonies or with simple multicellular organisms.

Studies of seabed deposits reported in 2014 reveal that the ocean was especially poor in trace metals at this time. The concentration of essential trace elements rose significantly about 660 million years ago, in time to produce the boost to evolution that was the Cambrian Explosion (see page 134). Interestingly, it also rose at the time of the GEO, another time when life flourished.

The low level of trace metals in the ocean can be explained by the geological inactivity that kept the supercontinents intact. The rate of tectonic activity we are familiar with today didn't begin until around 750 million years ago. The level of oxygen was also relatively low, especially in the deep oceans. In 1998, the American geologist Donald Canfield proposed that oxygen released in the GEO only oxygenated the top few centimetres of ocean water. Below that, high levels of hydrogen sulfide prevailed, washed there from the weathering of rocks and the oxidation of pyrite (iron (II) disulphide). Little, if anything, lived in the deep ocean.

Cell evolution

One important change took place during the Boring Billion years – the first eukaryotic cells evolved. These cells keep their genetic information coded in DNA locked away in a nucleus, bound with a membrane. Eukaryotes have other membrane-bound parts, too, called organelles, which perform specific functions. The archaea recorded in the early rocks were very simple types of cell, called prokaryotes. Prokaryotic cells have a long string of either DNA or RNA, but no nucleus to house it. They dominated the planet for about 2.5 billion years.

Eukaryotes probably first appeared around 1.5–1.84 billion years ago. 'Molecular clock' dating, which calculates from genome studies, gives a date of 1.84 billion years ago, while the oldest confirmed microfossils date from 1.5 billion years ago. Although the first eukaryotes were still single-celled organisms, all the later forms of complex life would develop from these biological pioneers – we are all eukaryotes.

Taken captive

Eukaryotes seem to have developed by enslaving prokaryotes. In 1967 American biologist Lynn Margulis argued that prokaryotes had been absorbed and put

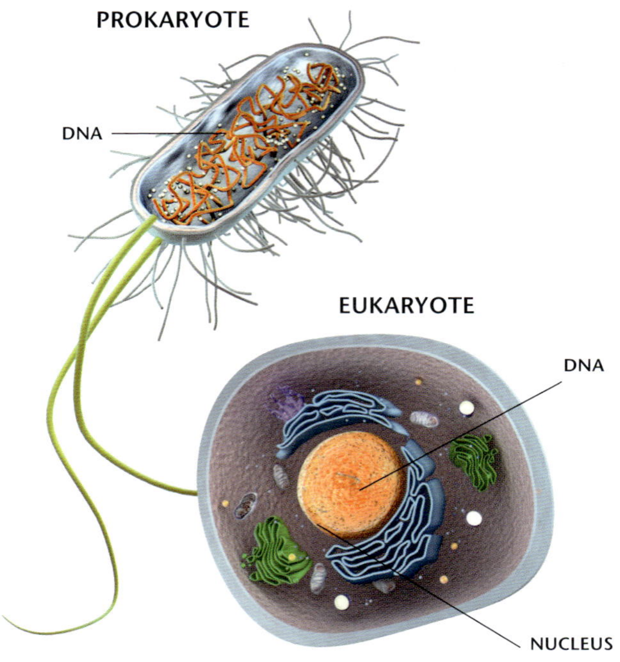

Prokaryotic cells are single-celled organisms. They have no nucleus; their DNA forms a free-floating loop in the cell.

Eukaryotic cells are more complex and can be found in single or multicelled organisms. They have a nucleus, which holds their DNA, and functional structures (organelles), each surrounded by a membrane.

to work within the eukaryotic cell in a process called endosymbiosis. This theory built on an idea first proposed in the 19th century, which had fallen out of favour. The chloroplasts in green plants, which carry out photosynthesis, and the mitochondria, which are a cell's 'powerhouse' and extract energy from chemical bonds, represent once-free prokaryotes now existing symbiotically within larger cells.

Eukaryotic cells can reproduce sexually, while prokaryotic cells only reproduce by dividing, with each cell an exact copy (clone) of its parent. Genetic variety comes about only through random mutation. If this is beneficial, the cell might survive and reproduce, perpetuating the mutation. Eukaryotic organisms can often combine DNA from two parents, making a pick-and-mix offspring with a random selection of genetic traits from each. Variety and beneficial traits can emerge much more quickly through sexual reproduction.

This might lead us to expect that eukaryotic cells would quickly gain the upper hand, but they faced a considerable hurdle. They need oxygen, and Earth was still oxygen poor.

Towards the end of the 'Boring Billion', more complex multicellular eukaryotic organisms finally began to develop perhaps as a result of increasing oxygen, increasing trace minerals and more diverse habitats. The first larger organisms may have been sponges, which probably appeared around 750 million years ago – just in time for some pretty cold weather with the start of the next Snowball Earth event.

The Earth moves

A giant barrel sponge spawning in the Sulu Sea, Philippines.

Tectonic activity was probably rapid in early Earth, forming about 70 per cent of the current rocky surface, then it slowed considerably from around three billion to 750 million years ago, as the crust cooled and thickened. Eventually, the upper mantle was cool enough and the crust was sufficiently thick for large slabs of oceanic crust to be pulled down into the subduction zones without breaking, drawing the seabed along and pulling at the mid-oceanic rifts. The continents began to move, and they haven't stopped since. The supercontinent Rodinia broke up and conditions around the world changed.

A supercontinent has a band of rainy land surrounding a parched interior. When the land is arranged in smaller blocks, with more coastline, there is far more rainy land and less dry land. Rain increases weathering of rocks, taking carbon dioxide from the atmosphere. This reduces the greenhouse effect and allows heat to escape into space, cooling the planet. The next Snowball Earth event struck around 715 million years ago, the temperature dropping to around -20 C (-4 F).

The planet might have been nudged towards this next icy catastrophe by a specific trigger, perhaps in the form of extreme volcanic eruptions. These could have filled the upper atmosphere with sulfur aerosols that reflect solar radiation back into space, preventing heat from the Sun reaching Earth. Eruptions such as this took place between 719 million and 717 million years ago, in an area that is now part of Arctic Canada. These, and the break-up of the supercontinent, may have pushed Earth to a tipping point, the temperature cascading downwards. Water froze out of the air and covered land and sea with a layer of frost, snow and ice; this further reflected solar radiation, intensifying the cold.

Living organisms probably played a part too. As organisms living near the ocean surface took in carbon dioxide, died and sank to the seabed, they carried their fixed carbon with them. This become incorporated into rock forming from the sediment and was

effectively removed by the carbon cycle. For 80 million years, the climate was in turmoil. An initial Snowball period lasted 58 million years. Then Earth thawed, but ten million years later it froze again.

Back from the deep freeze

As before, relentless volcanic activity would finally have produced enough carbon dioxide to re-melt the world. In 1992, the American geologist Joe Kirschvink suggested that there was a sudden transition to hot-house conditions at the end of a glaciation. The planet might have switched from a snowball to a greenhouse in just 2,000 years or so.

The relentless movement of glaciers during the snowball event grated away the surface of the rock and carried it into the sea as phosphorous-rich sediment. As the temperature rose and organisms began to reproduce more rapidly, there was a good supply of mineral nutrients to nourish them.

Life bounces back

There seems to be a link between Snowball Earth and the evolution of more complex organisms. The first non-microscopic animals – the Ediacaran biota – emerged 575 million years ago. They include the *Tribrachidium*, with its three-fold symmetry, and *Dickinsonia*, discovered in 1947. This flat, disk-like, segmented organism had no hard parts. It was finally classified as an animal – the earliest known – in 2018 after traces of molecular fossils of cholesterol, a biomarker for animal life, were found in one well-preserved specimen. As well as the hard-to-classify fossils of soft-bodied organisms, there are Ediacaran trace fossils such as worm burrows.

The Ediacaran wasn't a buzzing place. The seabed, at least in the shallows, was covered with a carpet of microbial slime, and on this lived strange quilted organisms either attached directly by stalks to the

***Left:** The Ediacaran* Dickinsonia *is the earliest known animal, from a period 575–541 million years ago.*

***Opposite:** A cast of the quilted* Charnia, *the first accepted complex Precambrian organism.*

sea floor or crawling along feeding on the slime. But soon that would all change. The seabed would become a hive of activity, sporting arthropods with jointed legs, the first creatures that could see, and fast moving predators with teeth.

An explosion of life

The Ediacaran period extends from the end of Snowball Earth, 635 million years ago, to the start of the Cambrian era, about 541 million years ago, when we might consider the modern world to have begun. The Cambrian saw rapid and extreme diversification of quite sophisticated animals – all the major modern animal phyla appeared during this time.

In the 19th century, Charles Darwin remarked on this sudden change in Earth's living community, finding to his consternation that it seemed to contradict his theory of evolution. He claimed that life had built up and changed slowly over millions of years. But the fossil record suggested an abrupt and lavishly populated starting point, with a great diversity of species appearing where before there had been nothing.

The geologist J. W. Salter suspected that there should be pre-Cambrian fossils in rocks taken from the Longmyndian Supergroup in Shropshire, England, but he found nothing except a few potential trace fossils of burrows. The same rocks have since been discovered to be rich in microbial fossils which the Victorian paleontologists could not see. The reason for 'Darwin's dilemma', as it is known, is that the pre-Cambrian organisms had no

> *'To the question of why we do not find rich fossiliferous deposits belonging to these . . . periods prior to the Cambrian system, I can give no satisfactory answer.'*
>
> Charles Darwin, *On the Origin of Species,* 1859

hard body parts so there was nothing that would readily fossilize.

Eat and be eaten

The Cambrian Explosion saw a gradual rise in oxygen levels, but a more important influence on evolution was the emergence of predatory behaviour. The Ediacaran organisms had fed on slime, the microbial mats and the drifting microbes of the sea – they had not eaten one another. When some of them started to do so, the relationship between predator and prey fuelled rapid evolution. Organisms that had been sessile and soft-bodied were an easy target, and had to evolve defensive strategies such as hard shells. To avoid being preyed upon, some organisms became capable of motion; and the advantage of sight, to both predator and prey, boosted the evolution of vision.

The effects of the Cambrian Explosion can be seen around the world, but a few particularly rich fossil beds, such as those in Canada (the Burgess Shale) and in China

TESTING THE CAMBRIAN CLIMATE

Scientists can examine ice cores and tree rings, including fossilized trees, to discover information about Earth's climate and atmosphere. But ice-core data only goes back a few million years. There is no ice-core data from before the start of the current glacial period, because there was no ice sheet in the warm interglacial period. There would have been no tree rings until trees evolved. To discover more about the Cambrian era, scientists measure the proportion of different isotopes of oxygen incorporated into fossils. Sea creatures fix oxygen from the sea in their shells, so the ratio of oxygen isotopes matches that of the ratio in the sea during their lives and can be used to calculate the temperature at the time. In 2018, analysis of microscopic shells from Cambrian brachiopods confirmed that the Cambrian Explosion happened during a greenhouse period, when Earth was warm.

The modern Priapulida *or penis worm is a simple, unsegmented worm that lives below the surface of the seabed. The Cambrian penis worm could turn its mouth inside out and use the tiny teeth in its cheese-grater-like throat to drag itself along the ground.*

(Qingjiang, discovered in 2007), have provided evidence of a considerable overlap of species.

The Burgess Shale

The Burgess Shale is an extraordinarily rich fossil deposit near Burgess Pass in the Canadian Rockies. It was discovered in 1909 by David and Helena Walcott, and was the first evidence of the Cambrian Explosion. Very diverse Cambrian fossils are extremely well preserved there, with soft body parts intact. There are ten separate pockets of fossil beds, with Walcott Quarry the most famous.

More than 200,000 fossils have been collected from the site, representing different types of arthropods, soft-bodied organisms and plants. There are also many microfossils of microbes and algae. Ninety-eight per cent of the fossils represent organisms with no hard body parts, which would usually be lost to fossilization. In total, around 150 species have been found.

The fossils were deposited around 505 million years ago, 35 million years after the Cambrian Explosion. They were probably preserved after a mudslide fell onto the area of seabed where the organisms lived, instantly burying them in deep sediment. They died instantly – their body positions show that they didn't curl up or try to burrow out of the mud. Canada and the location of the Burgess Shale were just south of the equator at the time, so the fossils represent the fauna of a tropical sea.

Around the same time as the Cambrian Explosion, another significant change took place. Organisms larger than algae and bacteria began to colonize an entirely new environment – land.

The Cambrian seas were populated by such strange creatures as these swimming Opabinia *– a jointed animal 5 cm (2 in) long with five eyes and a proboscis to grab prey.*

CHAPTER 7

Living **LAND**

'Keep steadily in mind that each organic being is striving to increase ... that each at some period of its life ... has to struggle for life, and to suffer great destruction. ... The war of nature is not incessant ... the vigorous, the healthy, and the happy survive and multiply.'

Charles Darwin, *On the Origin of Species*, 1859

For around three billion years, almost everything that lived on this planet, lived in the sea. Then, just a moment ago in geological time, pioneers crawled from the oceans and began to shape the world in new ways. Living things have made the Earth their own, changing rock, air and climate.

Tiktaalik *lived about 375 million years ago and grew up to 3 m (10 ft) long. It was discovered in 2004 and first described and recognized as a transitional creature in 2006. Although it had scales and gills like a fish it also had wrist joints and lungs and could support itself on land.*

The move to land

Apart from microbes, the first organisms to stray on to land made their move around 500–450 million years ago. There is a good chemical reason why life didn't venture on to land much earlier than this, and it is again linked to oxygen.

Water as sunscreen

Around 600 million years ago, the amount of oxygen in the atmosphere increased steadily and the planet developed a thin ozone layer. Ozone is produced when ultraviolet breaks apart some of the oxygen molecules in the atmosphere, then atomic (O) and molecular oxygen (O_2) combine to make ozone (O_3). As the ozone layer thickened, it blocked some of the ultraviolet from the Sun.

Ultraviolet is damaging to living cells. Water blocks ultraviolet, so organisms that live in the sea are protected from it. The Cambrian Explosion coincided with the ozone layer becoming thick enough to block some radiation, perhaps making it possible for organisms to live in shallower water. Photosynthetic organisms needed sufficient sunlight to penetrate the water to allow photosynthesis, but not enough to cause cell damage. Others were better off in deeper water, out of harm's way.

By around 480–460 million years ago, the ozone layer was thick enough for the first land colonists to survive out of water. They lived on rocks in the intertidal zones, where the water level rose and fell daily. From these, the first primitive land plants developed. They probably used chemicals such as scytonemin as a natural sunscreen. As the ozone layer thickened, this protection was no longer necessary and more organisms adapted to life on solid ground. But

Wiwaxia *was a soft-bodied Cambrian sea creature up to 5 cm (2 in) long. It was protected by spines along its back.*

before plants could exist, something else had to appear first – soil that would supply them with nutrients.

Making soil

Soil today is a complex mix of inorganic matter (sand, rock, clay) and organic matter (bits of dead plants, animals and microbes). The oldest samples of soil are around three billion years old, and are the result of the physical and chemical breakdown of rocks on Earth's surface. The first biotic soils developed as cyanobacteria and other microbes grew first in intertidal zones, and then moved further inland. As microbial mats began to green the land with slime, their tiny bodies added organic matter to the developing soil. Where this early soil collected in pockets on rocks near the shore, it didn't immediately provide a particularly hospitable environment, but became enriched over time. Even the simple cover of microbial mats stabilized the land to some degree, slowing the rate of erosion by holding the soil in place. The microbial cover would also have added organic acids to the soil. With matter held in place for longer, and with more chemicals to act on it, chemical weathering and decomposition could advance, but not yet far enough to support plants.

No trouble with lichen

The key to exploiting and making early soil more viable lay with lichen. As composite organisms, lichens are the product of a symbiotic relationship between algae or between cyanobacteria and fungi. The algae or cyanobacteria photosynthesize, producing energy; the fungi, with their long thin filaments, are adept at gathering water. Lichens containing cyanobacteria can take up nitrogen from the atmosphere and fix it, releasing it into the soil when they die. Most importantly, lichens can colonize bare rock – they don't need any soil. This equipped them to exploit the barren land, although we don't know exactly when they did this – perhaps some time between 700 and 550 million years ago.

Lichen growing on rocks is much the same now as it was hundreds of millions of years ago.

A HOLE IN THE OZONE LAYER

A layer of ozone 15–30 km (9–18 miles) above Earth's surface protects living things from harmful solar radiation. Ozone makes up less than 10 parts per million (ppm) or 0.001 per cent of the gases at that level, so the layer is not pure ozone.

In 1985, evidence emerged of ozone depletion over Antarctica; this became known as a 'hole' in the ozone layer. Investigation by NASA revealed that the problem extended over the whole of the region. Ozone depletion is not limited to Antarctica, but the effect is worse there than elsewhere.

Even before the discovery, two American chemists, Mario Molina and Sherwood Roland, had proposed in 1974 that CFCs (chlorofluorocarbon gases) could threaten the ozone layer. CFCs were widely used in aerosol cans and as coolants in refrigerators. The 'hole' forms each southern-hemisphere spring and partially heals later in the year, but it was healing a little less each year. In 1987, a worldwide agreement to phase out the use of CFCs was enshrined in the Montreal protocol, ratified by all 197 UN member countries. It was a remarkable act of cooperation in the face of looming climate catastrophe.

The ozone layer is now slowly recovering and the hole over Antarctica is expected to heal by 2060 if no further damage is done. The ozone crisis was the first hint that one species, humankind, could easily destroy the atmospheric conditions which make life on land possible.

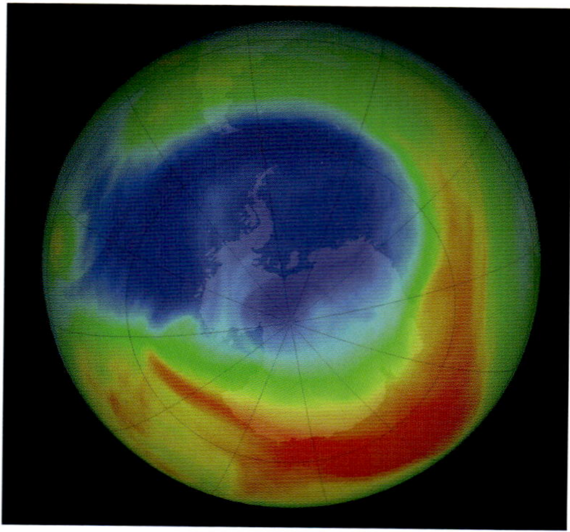

A false-colour version of a photo taken from space shows the area of depleted ozone over Antarctica in 2019.

As the lichen slowly moved further inland they made and enriched soil until a second wave of organisms could colonize the land. Simple plants evolved along the coast around 440 million years ago. At first, they were bryophytes – non-vascular plants such as liverworts and mosses. Like lichens, these also cooperated with fungi. Mycorrhizae, fungi that live among the roots of plants, evolved around 500 million years ago (before there was anything with roots). Living alongside the first plants, the filaments of mycorrhizae could burrow down into the rocks, releasing nutrients such as phosphorous and calcium and fixing nitrogen. The photosynthesizing plants provided the fungi with food in exchange.

Cooksonia, first discovered in 1937, is the oldest known plant to have a stem and vascular tissue. It's a transitional form between bryophytes and vascular plants.

The Foundation of Life

Mycorrhizal fungi helped to hold the particles of soil together and gave it stability. As the first plants died and decomposed, they added to the mix, making soil more complex, nutrient-rich and with a structure better adapted to retaining water. The ground was laid (literally) for the spread of plants on land.

These first simple plants provided food for animals. The first were invertebrates, moving on to the land in the middle of the Silurian (444–419 million years ago). They were animals such as mites, spiders and springtails which lived among the mosses and crawled over the lichen. Like the plants, they traced the path of water, living near rivers and streams.

Vascular plants with stems and, later, leaves evolved from bryophytes. They started to reproduce by making spores which could be blown further inland.

During the Devonian era (419–359 million years ago), more complex plants emerged. Their sophisticated root systems held the soil together much more firmly so it was not easy to wash or blow away. This newly stable material was supplied with a wider variety of chemical components and formed the thick, rich humus of the types we see now. Plants spread into the continental landmasses and areas of lush vegetation soon flourished in the warm climate.

'Slimy things did crawl with legs'

After plants, came animals. At first, these pioneering creatures were arthropods, but towards the end of the Devonian era, around 375 million years ago, they were followed by evolving fish which hauled themselves up the shore on fins that became prop-like with supportive bones. These 'fishapods' developed the ability to breathe air and evolved eventually into amphibians, the first four-legged land animals. Meanwhile, back in the sea, fish diversified rapidly; the Devonian era is sometimes called the 'age of fishes'. Among them were the lobe-finned fishes (from which terrestrial tetrapods evolved) and huge, armour-plated fish called placoderms.

As life on Earth developed into a complex ecosystem that extended further and further inland, animal waste was added to the soil, improving its fertility. During the Carboniferous (360–299 million years ago), vast forests spread across areas of hot swampland. They were populated by giant club mosses, tree ferns, horsetails and trees, and by prolific and ever-larger amphibians and arthropods. As the vegetation took carbon dioxide from the atmosphere and replaced it with oxygen, it raised the oxygen level to 35 per cent (today the atmosphere is just 21 per cent oxygen). The high oxygen content seems to have favoured outsized arthropods – dragonflies the size of seagulls

and centipedes two metres long, metre-long scorpions and cockroaches – as well as crocodile-like amphibians 6 metres (20 ft) long. This might sound the stuff of nightmares, but these creatures were the everyday reality of the Carboniferous forests.

The Carboniferous forest teemed with giant arthropods and amphibians.

CHANGING RIVERS

The growth of plants affects the landscape in a direct way. Where there is vegetation, the rapidly changing networks of braided rivers alter to a meandering, single channel; this is because plant roots keep the river banks in place and this directs the flow of water. One of the first signs of plant loss during an extinction event is the adoption by rivers of delta-like braided patterns.

Below: *The Rakaia River in New Zealand shows the braided pattern typical in areas with little vegetation.*

Bottom: *The Cononaco River in Ecuador has a single channel characteristic of rivers that cut through areas of heavy vegetation.*

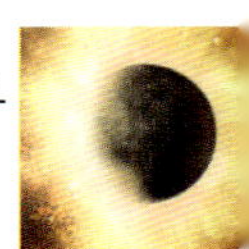

THE PRESERVATION GAME

The most familiar types of fossils are those of hard body parts, such as bones, teeth and shells, which fossilize most readily. They don't decay too quickly and often survive even if the soft parts of a body are consumed by scavengers or decomposers. Dinosaur bones and teeth and ammonite shells usually spring to mind when we think of fossils.

Soft tissue such as feathers and skin is sometimes fossilized if it is buried quickly enough in the right conditions. There are relatively few fossils of soft-bodied organisms such as jellyfish, not because they were rare but because they are not readily fossilized. Most species alive in the world today are insects, but few of them will become fossils. Any future species looking at the fossil record of the 21st century will probably gain a distorted impression of the distribution of hard- and soft-bodied species. Currently, half the biomass of animals is made up of arthropods, but butterflies, aphids and spiders are unlikely to become fossils. The chances are that we've been left a similarly distorted record of the past.

Trace fossils (or ichnofossils) preserve imprints of where organisms have been and include footprints, burrows, nests, excrement or stomach contents (coprolites). If an animal has made footprints in sand or mud that have been covered with sediment before fading away, it's possible to separate the different types of rock – the original ground surface and the compacted sediment – to uncover the footprints.

Another example of a trace fossil is the outline of soft tissue which has disappeared but left a cavity. Sometimes paleontologists can use trace fossils to piece together a dramatic scene from the distant past – the movements of one type of animal chasing another, or those of a parent walking alongside an infant, for example.

FROM TREES TO FOSSIL FUEL

When trees that spread all over the continental landmasses died and fell into the swamp, many were buried in mud and fossilized, eventually becoming the first coal deposits. The carbon dioxide taken in by these early trees was locked underground for more than 300 million years, leaving the animals of the time to bask in an oxygen-rich atmosphere. When the coal deposits were discovered and burned over the course of about 200 years, millions of years' worth of stored carbon was released into the atmosphere over a very brief period of time.

Much of the Carboniferous forest grew on the edge of tectonic plates, which would have been thrust together as the landmasses moved and the last supercontinent, Pangea, began to form. The deposits of undecayed wood were pushed underground, becoming

OIL AND GAS RESERVOIRS

Coal is made from plant material, but oil and natural gas (methane) are produced by the breakdown and compression of organic material on the seabed which originates with the deposits of algae, plankton and other organisms, most of them very tiny. When these organisms are buried deeply, the temperature and pressure rise; through a series of reactions, chemicals from their bodies are broken down and reformed as oil or gas. This migrates upwards through gaps in the rock until it meets a cap which it can't pass. A reservoir builds up beneath the cap over time, becoming the oil or gas deposits we exploit today.

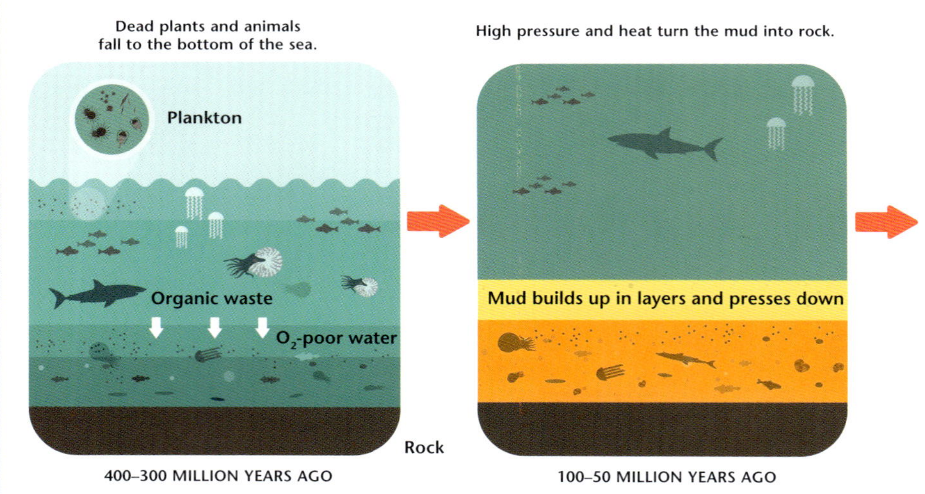

the coal seams that run abundantly through mountain ranges and hills today.

Animal, vegetable, mineral

Animals and plants of land and sea both fossilized, leaving us a record of their shapes and structures. Animals, particularly those with hard body parts such as shells, teeth and bones, have often been preserved intact. Although there are billions of fossils, fossilization itself was rare – most organisms broke down and their chemicals were recycled. Those fossils which have survived allow us to piece together the story of evolution on Earth.

'Sports of Nature'

In the past, people discovering fossils could not always work out what they were; they often considered them to be *lusus naturae* – 'sports of Nature'. When it was realized that some strangely shaped rocks are the

Much coal was laid down in the Carboniferous era, more than 300 million years ago, but oil is generally more recent. The oldest datable oil is only about 200 million years old and most is a lot younger.

The formation of oil and gas from 400 million years ago to the present day.

remnants of living things, it revolutionized our understanding of Earth and its history. Today, the notion is so familiar to us as to be unremarkable.

As long ago as the 6th century BC, the Greek philosopher Xenophanes of Colophon (*c*.570–*c*.478 BC) recognized the fossilized shells of molluscs as remnants of creatures that had died long ago. He deduced that the dry land in which they were found had once been a seabed, and used this to support his theory that everything is made of earth and water. He also surmised that Earth goes through dry and wet phases, and during wet phases everything turns to mud and all humans die. The fossils of sea creatures showed that they had lived in a wet phase of Earth's history. Although he was wrong, Xenophanes was the first person to use fossil evidence in a para-scientific argument. Herodotus (484–425 BC) cited seashells found in Egypt as evidence

'Fossils have long been studied as great curiosities, collected with great pains, treasured with great care . . . and this has been done by thousands who have never paid the least regard to that wonderful order and regularity with which Nature has disposed of these singular productions, and assigned to each class its peculiar stratum.'

William 'Strata' Smith, English geologist, 1796

Left: *A trace fossil – a dinosaur footprint from the Sataplia region of Georgia.*

Facing page: *The Taihang Mountains in Shanxi, Hebei and Henan provinces in China were once underwater, and yield marine fossils.*

that the land had once been underwater. He wrote that he saw in a valley in the Mokattam mountains 'the backbones and ribs of such serpents as it is impossible to describe: of the ribs there were a multitude of heaps'. Eratosthenes (276–194 BC) and Strabo (64 BC–AD 24) wrote about marine fossils, indicating that land now above the sea had once been underwater.

Around 1,500 years after Xenophanes, and the other side of the world, the Chinese administrator and scientist Shen Kuo (1031–95) concluded from the presence of fossilized marine creatures that the Taihang mountains had once been beneath the sea. He also realized that a fossilized bamboo forest in an area no longer suited to growing bamboo was evidence of climate change.

Leonardo da Vinci realized that the fossilized seashells he found were the relics of long-dead organisms. (He didn't publish his findings, but kept them hidden in coded notebooks.) These, like the seashells found by earlier writers, closely resembled extant species so were quite easy to identify.

More challenging were the fossils which didn't resemble extant organisms. Some early scholars took these as evidence of the existence of mythological beasts, including dragons. A Christian interpretation had it

that these long-lost creatures were killed off in Noah's flood and never seen again; the flood was used to explain why some fossilized sea creatures ended up on high ground – they had simply been stranded by the receding floodwaters.

> *'I saw bivalve shells and ovoid rocks running horizontally through a cliff like a belt. This was once a seashore, although the sea is now hundreds of miles east. What we call our continent is an inundation of silt.'*
>
> Shen Kuo, *Dream Pool Essays*, between 1080 and 1088

Grown from the ground

The way in which fossils formed was puzzling. Aristotle had suggested that they grew naturally in the ground to look like organic forms. It was often suggested that the force which shaped plants and animals was the same as the one which moulded stones into shapes that resemble natural organisms. The correspondence between shapes was thought to be a result of how that force works – or could indicate the greater correspondences between the micro and macro realms that were to be expected in a harmonious universe perfectly ordered by a god.

A neo-Aristotelian explanation suggested that a vegetative spirit (*anima vegetativa*), thought to produce the spontaneous

WHERE THE SEA HAS BEEN

The Chinese polymath and diplomat Shen Kuo's *Dream Pool Essays* (1088) recorded the natural phenomena and wildlife he observed and investigated on his travels around China. He outlined a theory of rock formation which involved the erosion of mountains, the deposition of silt to form sedimentary rocks, and the action of uplift to create mountains. He identified fossils of seashells far inland and interpreted these as evidence that the mountains had once been under the sea. These discoveries predated their European counterparts by 500 years or more.

generation of some organisms directly from inanimate matter, could also act on rocks to produce similar forms. Therefore, a fish-like fossil was the product of the same forces (acting on rock) as those that acted on organic matter to generate a fish, perhaps from the mud at the bottom of a river.

In 1027, the Persian naturalist Ibn Sina (known in Europe as Avicenna) developed an idea first proposed by Aristotle that some kind of 'petrifying fluid' was responsible for turning ancient shells to stone. Albert of Saxony expanded further on the idea in the 14th century, and it went unchallenged for at least another 200 years. A French potter, hydraulics engineer and amateur natural scientist, Bernard Palissy (1510–89) proposed that minerals dissolve in water to form 'congelative water'; they then precipitate, petrifying dead organisms and creating fossils. This account is not so very far from the truth.

In 1546, Agricola described various types of stones which he believed to resemble living organisms. He didn't suggest that they were organic, preferring the traditional account that they had grown in the ground and adopted organic shapes, but were not of organic origin. ('Fossils', at this time, meant anything dug up from the ground.)

> *'Between one layer and the other there remain traces of the worms that crept between them when they had not yet dried. All the sea mud still contains shells, and the shells are petrified together with the mud.'*
>
> Leonardo da Vinci, *Codex Leicester*

With the renewed and more enlightened attitude to science that characterized the later years of the 17th and 18th centuries, the debate about whether fossils were of organic or inorganic origin heated up. Scientists such as the English microscopist Robert Hooke (1635–1703) and the Danish physician and geologist Nicolas Steno argued that fossils formed from previously living organisms which had somehow become petrified. On the other side of the debate, the English naturalist Martin Lister (1639–1712) and the Welsh naturalist Edward Lhwyd (1660–1709) argued that

For some early thinkers, Noah's flood seemed to explain why marine fossils are found on mountaintops.

fossils were quaintly formed rocks which had never been living things. Lhwyd championed the idea that fossils grew when seeds from living organisms were washed into rocks. The same generative force that caused them to grow into organic beings caused them to grow, in this environment, into fossils.

FROM THE FISH'S MOUTH

In 1666, two fishermen caught an enormous shark off the coast of Livorno in Italy. The fish's head was sent to Steno who was working in Florence at the time. He noticed that its teeth bore a curious resemblance to a type of stone known at the time as 'tongue stones'. Known for centuries, tongue stones had been found in the rocks of Italy, even far from the coast. Steno deduced that they were, in fact, the teeth of some long-dead animal which had somehow been turned to stone.

The similarity between tongue stones and the teeth of extant sharks was sufficiently obvious to have been noticed before. In 1616, the Italian naturalist Fabio Colonna had stated that tongue stones were sharks' teeth. Contemporaries of Steno, such as Hooke and the English naturalist John Ray (1627–1705) came to similar conclusions, but Steno took the idea further, developing a consistent scheme from it and considering how fossilization might occur. At the time, the idea that all matter is made of 'corpuscles' (what we would call atoms and molecules) was gaining ground. Steno supposed that, over a period of time, the corpuscles in the original teeth would gradually be replaced by corpuscles of minerals and would slowly change into rock.

FOSSILIZATION REVEALED

Today, two processes for making body fossils are recognized: permineralization and replacement. Permineralization occurs when ground water carrying dissolved minerals percolates through tiny pores in material such as bone, shell or wood. The minerals are deposited in the cavities, adding to the structure, while the original material remains largely intact. Fossilized dinosaur bones and wood are often made in this way. The replacement process dissolves the original material and replaces it with minerals. The minerals that most often replace others are silica, pyrite and hematite.

The process of fossilization can only begin in the right circumstances. Generally, the organism must be rapidly buried before its body is torn apart by scavengers, scattered by the wind or otherwise destroyed. Organisms that live in areas where rapid burial and steady deposition of sediment are common (those living in or near water or mud) are more likely to be preserved as fossils.

Palissy's biologically accurate ceramics are testament to his extraordinary talent. In addition to explaining fossilization, he uncovered the origins of springs, gave a theory of earthquakes and volcanoes, designed gardens for the Tuileries, Paris, and described systems for delivering clean water. He died in prison at the Bastille at the age of 80.

Digging out the truth

Steno studied the rocks and cliffs of Italy in search of an explanation for how the sharks' teeth could have become embedded in rock that is not underwater. He noticed that rocks often show distinct layers, and devised his rules of stratigraphy (see page 81) and a theory to explain the placement of fossils. He suggested that all minerals were originally fluid and eventually settled, probably out of the ocean, forming layers. As later layers were laid down, they would be nearer the surface, unless something had disrupted them. Steno supposed that sometimes animals (or their dead bodies) could be trapped as the rock was laid down. This explained why fossils were often found deep within rocks. He noticed, too, that the oldest rocks contained no fossils, whereas younger rocks often

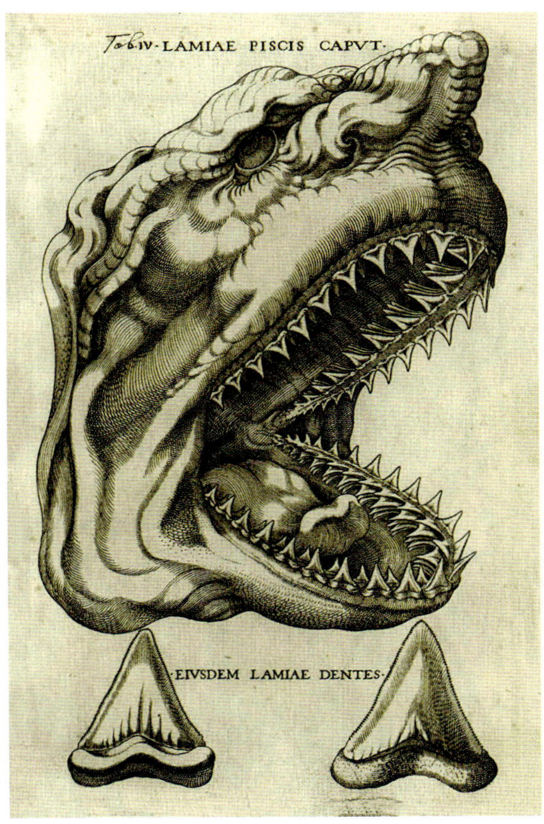

Steno's fearsome drawing of a shark's head and teeth, which he recognized as being similar to 'tongue stones'.

held abundant fossils. He concluded that the oldest rocks had been laid down before there were living animals; the fossiliferous layers, however, had been laid down by rock precipitating from Noah's flood (and later).

The problem of extinction

An increasing number of fossils were uncovered which looked approximately but not exactly like plants or animals. If they represented organisms, they were of organisms no longer around, and this didn't fit with the contemporary view that God had populated Earth in one great creative binge. The fact that they were similar rather than identical to existing organisms made it less likely that they were of organic origin.

John Ray concluded that while some fossils might have had organic origins, not all of them did. Ammonites presented a particular problem for him: they are so abundant and widespread, yet not a single related species had ever been found. Ray felt compelled to regard ammonites as of non-organic origin.

Laid in layers

The first geologist to realize that fossils could offer a way of dating rock was the English surveyor William Smith (1769–1839). Like Steno, he recognized that rocks are laid down in layers and always appear in the same order. But he made an additional important discovery – that particular types of fossil are found in particular strata.

Smith worked as a surveyor in coal mines, helping landlords to locate building stone and coal, and later working on canal cuttings. He saw at first hand how the same strata of rock are found across England, Wales and Scotland. In 1799, Smith, Joseph Townsend and Benjamin Richardson (the latter both clergymen with an interest in fossils) made a chart of the geological strata found near Bath and named them. They noted physical characteristics of the rocks and the types of fossils present in them. It

was the start of the work that would end in Smith's comprehensive and beautiful geological map of Britain, published in 1815 – the first national geological map anywhere in the world.

Over 15 years, Smith single-handedly mapped the geology of England, Wales and southern Scotland, covering an area of more than 175,000 km^2 (67,500 square miles). He wanted his map to be easy to understand as well as visually pleasing, and hit on the idea of using different colours for each of the strata. The maps are hand-coloured in 23 different tints. The base of each stratum is in the darkest shade, becoming paler further up so it's easy to see the order.

Evolving before evolution

Smith's work on fossils predates Charles Darwin's theory of evolution, published in 1859, yet strongly suggests it. The scheme of biostratigraphy relies on the notion that organisms change over time, that some die out and others appear. Indeed, the most useful organisms for the purposes of geological dating are those which are common and widespread but for a relatively brief period of time. When Darwin came to his work on evolution, the fossil record provided him with important evidence. As more and more fossils were discovered, and people studied them in greater detail, the notion of extinction became compelling.

> *'Either these [fossils] were terriginous, or if otherwise, the animals they so exactly represent have become extinct.'*
>
> Martin Lister, 1678

That the fossil record supports evolutionary ideas was not universally accepted, however. Evolution was not invented by Darwin – it was already a topic under discussion in the 18th century. The clash with religious thinking was already underway, with huge antipathy towards the idea that God might have created some creatures which he then dispensed with or, worse still, 'improved'. The pioneering French paleontologist Georges Cuvier was famously against evolutionary theories, though he accepted extinction.

Cuvier, a contemporary of Smith, began his professional life comparing fossils with extant species. He was the first to establish that Indian and African elephants are different species and also that they

differ from the extinct mammoth and an American fossil dubbed the 'Ohio animal' (which he later named the mastodon). He noted the link between living sloths and a large South American fossil which he named megatherium (now known as an extinct giant ground sloth). His publication on these two in 1796 effectively settled a long-running debate about whether organisms ever become extinct – they clearly do, and it would be hard to hide megatherium, even in 18th-century Paraguay.

Fighting change

Cuvier was adamant that organisms do not change gradually over time. His evidence for this included the mummified animals which Napoleon's expeditionary troops brought back from Egypt. These had been mummified thousands of years previously, yet were identical to extant species. When Jean-Baptiste Lamarck (a proponent of evolution) argued that evolution was too slow to show up in so brief a period, Cuvier said that if there is no change at all over a short period, there can be no change over a long period either.

Although he didn't believe in evolution, Cuvier saw that new organisms appear in the fossil record. His preferred explanation was that Earth lurches from one catastrophe

Facing page: *Ammonites, such as the one being uncovered here, are extremely common fossils.*

Right: *Smith's national geological map of Britain, the first of its kind.*

WILLIAM SMITH (1769–1839)

William Smith's father, a blacksmith, died when his son was only eight years old. Smith was sent to live with his uncle, who had a farm. There he collected the 'poundstones' that milkmaids used to weigh butter; these were, in fact, uniformly sized fossilized sea urchins. He also played at marbles using 'pundibs', which were fossilized brachiopods (sea creatures similar to clams).

At the age of 18, Smith began work as a surveyor. The Industrial Revolution was in full swing in England, and the nation's appetite for coal to power its new machinery was prodigious. Smith spent a lot of time examining mines, and soon noticed that as he descended a mineshaft he could see different layers of rock which followed predictable sequences from one mine to another.

William Smith's work as a surveyor exposed him to the rock formations and layers around the country.

He soon had the chance to test his suspicion that the same pattern of strata might be found throughout the country. The coal not only had to be found and dug out, but also moved around. A network of canals was built to ferry goods across the country. Smith began work on the Somerset Coal Canal, surveying in 1794 and observing excavation from 1795. Soon he was travelling on foot, on horseback and by carriage, covering around 16,000 km (10,000 miles) a year, surveying canal routes. As canals are cut straight down, this gave him a perfect opportunity to compare the strata revealed across the land.

Smith found that while the predictable sequence was found everywhere, it was not always possible to identify a stratum just by the appearance of the rock. He soon found that fossils held the key to this. While some fossils are found across several layers, others only appear in a single layer and can identify it uniquely.

He accumulated a large collection of index fossils which he used to identify layers. His principle of 'faunal succession' (the identification of rock strata according to fossilized flora and fauna) is still used in geology today, and he is credited with starting the field of biostratigraphy (dating rock strata by fossils).

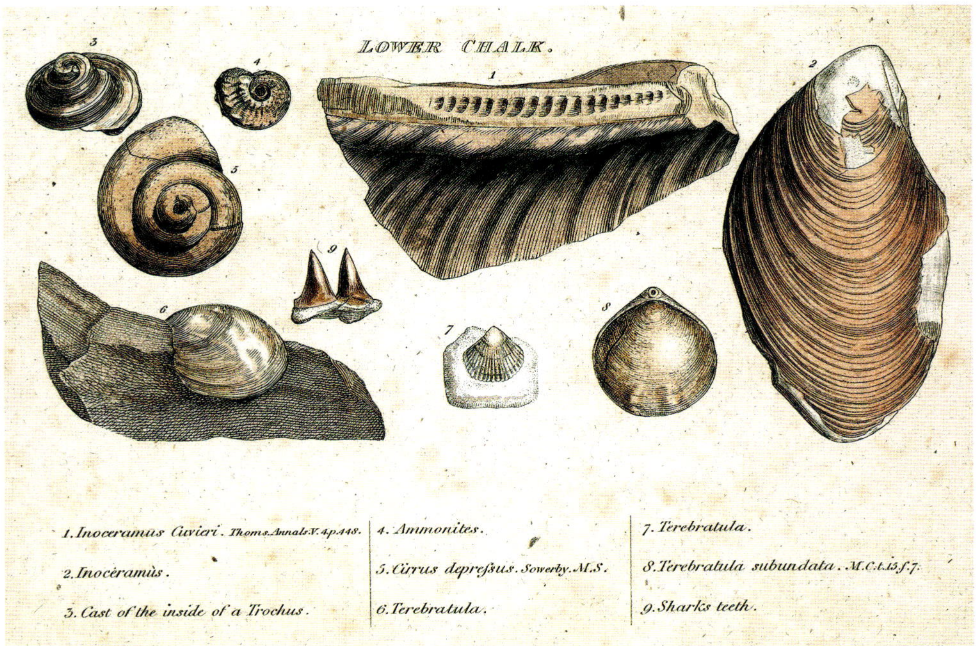

William Smith's illustrations of some of the index fossils which he used to identify the rock strata he examined in different areas.

to another; each catastrophe wipes out a clutch of organisms, then new ones appear in their place (though Cuvier didn't suggest a mechanism by which they appear). The most recent catastrophe in this scheme was a great flood. Cuvier cited authorities among the Ancient Greeks who wrote of such a flood, and indigenous North American tales that told of a deluge.

Catastrophe!

Cuvier's notion of periods of extinction was the basis for a theory of catastrophism which became a dominant model for the history of Earth in the early 19th century. Between catastrophes, Cuvier believed that life on Earth was fairly stable, with organisms staying the same. This pattern would have had to occur over a long period of time, leading Cuvier to the conclusion that Earth must be several million years old. This was at odds with the conventional belief, supported by the Christian Church, that only a few thousand years had passed since Creation.

Although Cuvier proposed a flood as the most recent catastrophe to have wrought change on Earth, he did not link it with Noah's flood. Unfortunately, when the English geologist William Buckland translated Cuvier's work on fossils, he added an introduction making that link. Cuvier had suggested a localized flood of extended duration, while the flood of the Noah story was supposedly global and

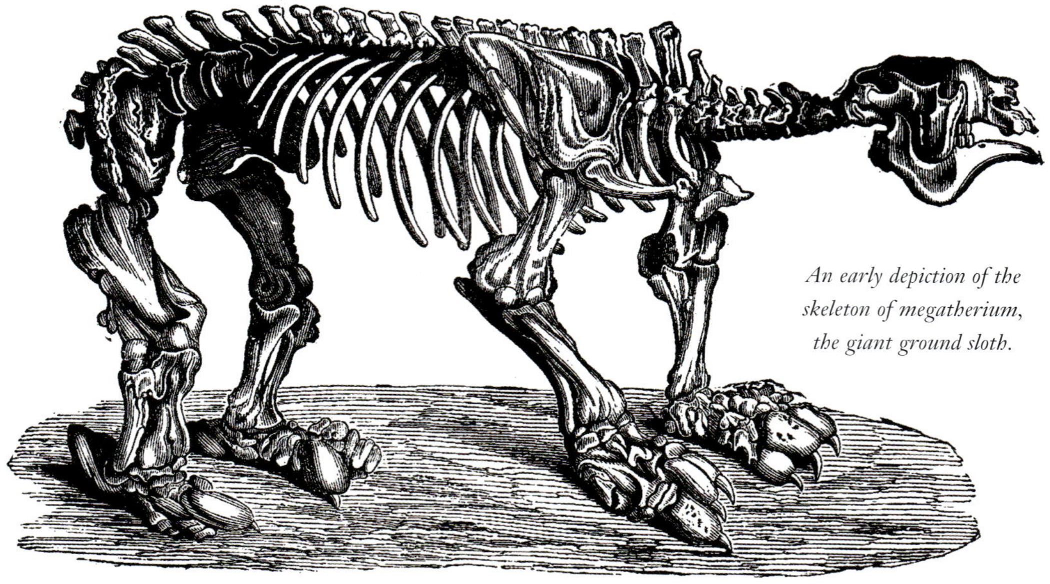

An early depiction of the skeleton of megatherium, the giant ground sloth.

of short duration, but that didn't trouble Buckland who persisted in trying to find geological evidence for Noah's flood. This misrepresentation of Cuvier's work meant its reception in Britain was tinged with a religious aspect he had not intended and which did it no good. Catastrophism became unhelpfully associated with a religious view that interpreted natural events as episodes of divine intervention. Buckland eventually gave up his preoccupation with the flood to investigate a possible glaciation catastrophe, the European ice age first suggested by Goethe in 1784 (see page 126).

SLOW AND STEADY

The opposite view to catastrophism was slow change. In the case of organisms, this was evolution. The first proponent of evolutionary theory was Jean-Baptiste

ELEPHANTS FROM THE PAST

The mastodon teeth which Cuvier examined and declared to be different from extant elephants had been uncovered by slaves working a plantation in South Carolina in 1725. As they were African-born, the slaves were familiar with elephants and could identify the strange, large, ribbed stones as bearing a striking resemblance to elephant teeth. These were the first documented North American fossils.

Lamarck, a French zoologist who had been put in charge of (and made great progress in) invertebrate zoology at the Musée National d'Histoire Naturelle in Paris.

Although Lamarckian evolutionary theory is frequently mocked today, it is a coherent explanation of how animals evolve. According to Lamarck, organisms change in response to change in their environment and the consequent demands placed upon them. At first, behaviour alters and as a result the bodies of animals adapt over generations. The first 'law' of this hypothesis is that use or disuse drives the development of structure: use encourages growth and disuse leads to contraction or loss (so moles have small, weak eyes because they don't use them). The second law is that the changes wrought by use or disuse are hereditable. Lamarck's often-cited example of the giraffe makes the reasoning clear. Originally, short-necked giraffes stretched up to reach high-growing leaves. A 'nervous fluid' flowed into the neck, making it grow. This continued over generations, with the offspring inheriting the stretched neck, and then stretching it further, until the modern giraffe form was achieved.

Lamarck also believed that evolution is directed, with Nature working towards perfection and complexity in organisms. He suggested that simple organisms such as protists were spontaneously generated continuously. He didn't believe that organisms ever went extinct, but changed form by evolving. Religious and scientific opponents denigrated this theory. For the religious, the notion that the natural world is not the realization of God's perfect plan but the product of blind forces was repellant. The scientists considered the argument insufficiently rigorous. Lamarck died in poverty and obscurity in 1829.

> *'All of these facts, consistent among themselves, and not opposed by any report, seem to me to prove the existence of a world previous to ours, destroyed by some kind of catastrophe.'*
>
> Georges Cuvier, 1796

Georges Cuvier was an early pioneer in arguing for an old Earth, based on his catastrophist model of widely spaced periods of sudden change.

Lamarck was not the only person to be considering evolution as a possibility. Darwin's own grandfather, Erasmus Darwin, outlined a theory somewhat similar to that of Lamarck in *Zoonomia, or, The Laws of Organic Life* (1794–96). He proposed that life evolved from 'one living filament', but struggled to explain how one species could

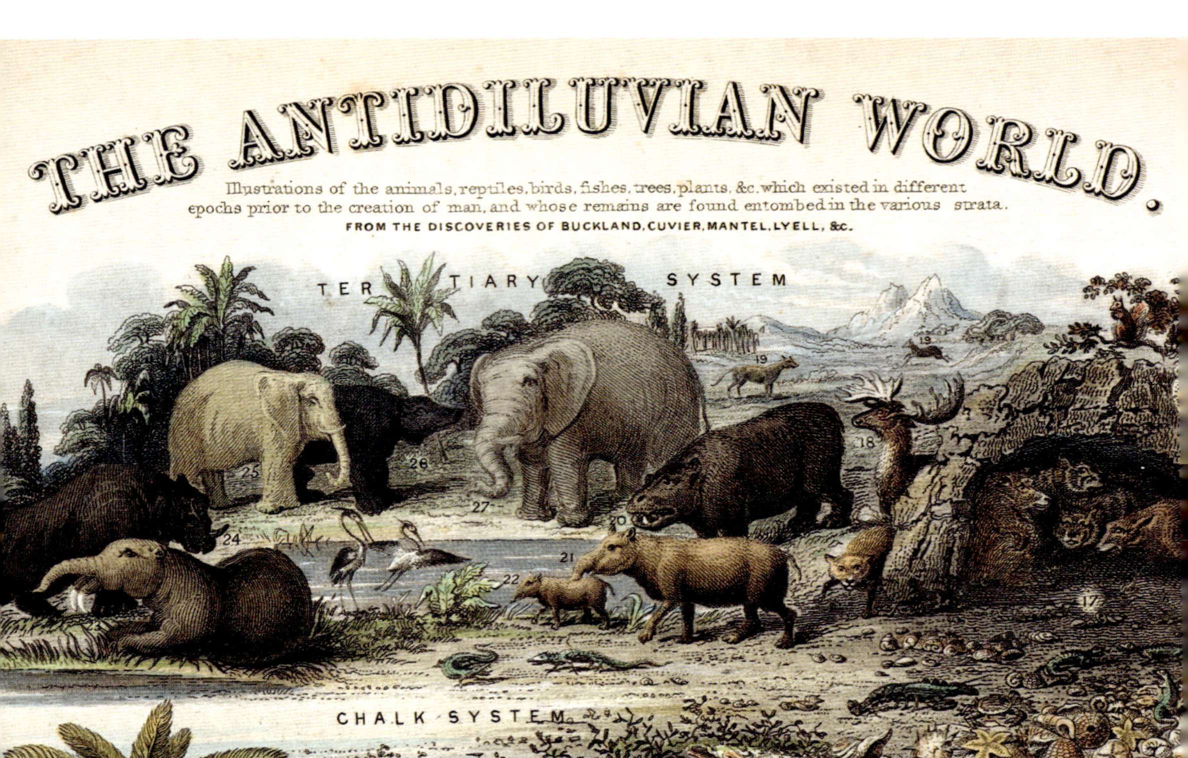

The notion that different species existed before a catastrophic flood was not fully exploited by this artist, who drew animals rather similar to those still extant in the 19th century.

evolve into another. He edged towards sexual selection in saying that 'the strongest and most active animal should propagate the species which should thus be improved'.

How catastrophic is a catastrophe?

As we have seen, in 1785 James Hutton proposed that slow processes produce change; these processes are not simply ongoing but are the same now as they have ever been. This model has no space or need for rapid catastrophes that wreak havoc in the geological environment.

The fossil record, though, seems to offer a challenge to uniformitarianism. Careful examination of the record turns up several events during which entire classes of fossils disappear over what appears to be a short time. These are not just extinctions, but mass extinctions. The most famous, though not the most extreme, was that at the end of the Cretaceous period (the KT extinction event) which removed all the non-bird dinosaurs from the land, the pterosaurs from the sky and all the ammonites and large reptiles from the seas. This appears

to be the very definition of 'catastrophic'.

Fast and slow change

By the end of the 18th century, following Smith's use of fossils for identifying and ordering strata, fossils had become integral to geology. Cuvier conducted work in the Paris basin which demonstrated, like Smith's work in England, that different fossils are associated with different strata. His explanation relied on catastrophes to mark the shift between the conditions and creatures of each age. These changes were apparently drastic, as the area around Paris had been sometimes under the sea, and sometimes flooded with fresh water.

William Buckland (see pages 157–8), who had given a religious angle to Cuvier's work, was teaching at Oxford when the English lawyer Charles Lyell decided to turn his hand and mind to geology. Lyell found Buckland's catastrophism unsatisfactory, taking particular issue with the attempt to link geology with Noah's flood and giving God control of the formation and history of Earth. Lyell was determined to build geology as a respectable science based on sound empirical evidence.

He re-examined the geology of the Paris basin and investigated many other geological sites around Europe. Taking his cue from the rather neglected work of Hutton, who was convinced that geological

Above: *Jean-Baptiste Lamarck, whose ideas have often been ridiculed. But recent developments in epigenetics now find that some heritable changes don't require alterations to DNA. This corresponds with Lamarck's idea about the action of 'soft inheritance'.*

Left: *Vision is not necessary for the mole's underground life; it might eventually lose this sense altogether.*

processes are slow, Lyell found that he could explain geological change without recourse either to God's hand or sudden catastrophic events. Using Hutton's cycle of erosion, deposition, compaction, heating and uplift, Lyell argued that geological processes are uniform through time and he formulated the doctrine of uniformitarianism (a term coined by William Whewell – who also coined the word 'scientist'). Change happens around us, but it is so slow that we don't notice it.

In the Paris basin, Lyell didn't find evidence of sudden, cataclysmic change, but proof that Earth undergoes regular cyclical changes over an extremely long timescale. The extinctions and changes in the rocks only appear to be sudden because millions of years are represented in just a few centimetres of deposited sediment and fossils. The city of Paris was once below sea level, but its altered landlocked location was not, Lyell argued, the result of cataclysmic events – the change was likely to have occurred over very long periods of time.

The Principles of Geology

Lyell published his seminal work, *The Principles of Geology*, in 1830. Its central thesis is that the surface of the Earth is the product of millions of years of small changes produced by natural processes. Charles Darwin read Lyell's book during his journey aboard the *Beagle*. In Valdivia, Chile, Darwin experienced a terrifying earthquake and heard there had been volcanic eruptions nearby. He concluded that the same force was behind both, and that volcanoes were linked by magma underground. He suggested that if the pressure inside volcanoes was not released, it could cause earthquakes. Darwin was so convinced by Lyell's hypothesis that when he came to write *On the Origin of Species* he based it on the same principle of gradual cumulative change. But where Lyell had investigated geological change, Darwin focused on the driving forces of reproduction, competition and inheritance among animals.

Mural mussels

Lyell later reviewed his original thesis to allow for more rapid transformations in some circumstances. While visiting the Roman Temple of Serapis in Pozzuoli, near Naples, Italy, he noticed three tall, stone columns which had a line around them some distance off the ground; he recognized that these lines had been made by a type of mussel. It occurred to Lyell that, at some point after the Romans had built the temple, the columns must have been underwater. The sea level had risen and fallen again in the intervening 2,000 years. Lyell realized that the processes were acting relatively rapidly in terms of geological time. The earlier shoreline had been 2.74 m (9 ft) higher than it was in Lyell's time, but it's since risen a further 3.15 m (10.3 ft), so is slightly higher now than it was when the temple was built. Geologists believe this is the result of magma beneath the ground swelling and receding. This explanation accounts for the relatively rapid (in geological terms) changes.

Deep time and the rock cycle

Lyell's hypothesis introduced the concept of 'deep time' – a history of inconceivably great length. If mountains are constantly

CHARLES LYELL, 1797–1875

Charles Lyell was born the oldest of ten children in Kinnordy, Scotland, where his father was a botanist and a literary translator. The family had a second home in the New Forest, Hampshire.

Lyell studied at Oxford, where he attended William Buckland's lectures on geology, hearing his Christianized version of Cuvier's catastrophist theory of Earth's evolution. In 1820, Lyell became a lawyer, joining the bar in 1825, but realized that his true interests lay elsewhere. He had continued to pursue geology, and as his eyesight began to fail he gave up law in 1827 to become a geologist. During the 1830s, he was professor of geology at King's College, University of London, and published the first volume of *Principles of Geology*. A synthesis of Hutton's ideas and Lyell's own observations and deductions, it is considered one of the most important scientific books ever published. A later book, *Geological Evidences of the Antiquity of Man*, published in 1863, presented evidence that humans had been on Earth for a very long time but did not unequivocally endorse the theory of evolution. Lyell remained convinced that humans are somehow special in the natural scheme.

A close friend of Charles Darwin, Lyell encouraged him to publish his theory of evolution, even though his own religious beliefs meant he had reservations about the idea. At one point, he believed that the appearance of different organisms in different regions had come about through local centres of creation; he was never convinced that new species could arise by entirely natural processes.

The geologist Charles Lyell was knighted in 1848 and made a baronet (a hereditary honour) in 1864.

At the Temple of Serapis, Lyell (the seated figure on the left) noticed that the water level had dropped and risen considerably since Roman times – an indication of considerable geological change.

being eroded and yet we see no difference in our lifetime nor any discernible change even by referring to our oldest records – the accounts of the Ancient Greeks, for example – then the process must be very slow indeed. If a mountain grows or shrinks steadily by an average of a centimetre a year, it would take 10,000 years (twice the entire span of human civilization) to grow or shrink by just 100 m (328 ft). This change would be impossible to detect without advanced technology to measure it. And what unconscionable length of time would it take to raise mountains from a plain or reduce them to dust?

When it became obvious that slow processes were effecting geological change, it was impossible to accept the conventional view that Earth was only a few thousand years old. Even Kelvin's notion that Earth might be 100 million years old fell far short of the mark.

Around and around

Today we know that cyclical processes of formation and destruction together form the rock cycle. Hutton was the first person to propose a cyclical model for geological processes, though his cycle, predating knowledge of tectonics, was considerably simpler. It focused on heat acting to transform sediment into rocks and produce uplift (see page 85). The modern version of the rock cycle is more detailed.

Changing world

With life flourishing on land and in the oceans, Earth was subject to more changing influences than when it was a cooling lump of barren rock. The organisms living and dying on Earth altered the soil, the rocks, the air and the water, and the changes they wrought transformed life. The atmosphere, hydrosphere and lithosphere were joined and moulded by the biosphere; over the next few hundred million years, their fates would become ever more closely entangled.

Mount Etna, on the island of Sicily, as shown on the frontispiece of Lyell's Principles of Geology, *volume 2, in 1832.*

CHAPTER 8

Days of the DEAD

'The Dodo used to walk around,
And take the sun and air.
The sun yet warms his native ground –
The Dodo is not there!'

Hilaire Belloc, 'The Dodo', *The Bad Child's Book of Beasts*, 1896

No sooner had nature embarked on the great adventure of colonizing the entire land than it ran into trouble – death on a massive scale and in a repeating cycle. Life on Earth fell into a pattern of diversification followed by extinction, followed by diversification in new directions. This cycle is an intricate interweaving of geology, meteorology and biology.

The dodo, the most famous recent victim of extinction, roamed the forests of Mauritius until the 17th century.

Fishapods and tetrapods

Soil enabled plants and animals to move out of the oceans. Fishapods gave rise to amphibians, which still laid their eggs in water, but as adults were able to breathe air and live at least some of their time on land.

Amphibians, like these toads, live on land as adults, breathing air, but must keep moist and lay their eggs in water.

Mind the gap

We can only work out how organisms developed by investigating their fossils. But sometimes there are no fossils to examine. These gaps in the fossil record mean that scientists have little or nothing to go on. Frustratingly, the movement of creatures from the sea to dry land falls in such a gap. It's not clear why this should be – until recently it was thought that perhaps geological conditions didn't favour fossilization or there was little biological diversity during these gaps. Maybe geologists simply hadn't been looking in the right places. Generally, fossils are only uncovered when they are near the surface of the Earth or at coasts where large areas of rock are exposed. There are probably a great many fossils that will never come to light because they are buried deep underground or beneath the seabed.

The gap in the fossil record that occurs around the time of the colonization of land by tetrapods (animals with four feet) is known as Romer's Gap, after Alfred Romer who identified it. Romer's Gap falls 360–345 million years ago, at the end of the Devonian period and the beginning of the Carboniferous. The Devonian ended with a mass extinction event even more damaging than that which ended the reign of the non-bird dinosaurs. Following this event, sharks and ray-finned fish swam in the seas and amphibians walked on the land, but there is scant evidence of the change from fishapod to amphibian.

The fishapods seem to have made their move before the extinction event, flourishing during Romer's Gap and emerging as amphibians by its end. In Scotland and Nova Scotia, fossils have emerged which suggest that amphibians were diversifying during this period (Scotland and North America were a contiguous landmass at the time). Evidence from Scotland also undermines an earlier theory that low levels of oxygen during the Gap might have led to low diversity. Among the important Scottish discoveries is *Pederpes*, now considered the first modern tetrapod. It had forward-facing front feet with five digits, and a high, narrow skull that might indicate it breathed using muscular action rather than with the throat-pouch pumping mechanism of frogs.

A fossil of the tetrapod Pederpes *was found in Scotland in 1971, but was not recognized as an amphibian until 2002.*

Wet bodies

The lush, warm Devonian period was an ideal environment for amphibians, which is why they evolved as they did, since climate and living organisms are tied together. Extensive inland swamps meant they had easy access to water. The growing arthropod population gave them something to eat. Arthropods such as dragonflies, which lay their eggs in water and live there as nymphs, provided amphibians with a supply of food both in water and on land, so feeding both adult and immature amphibians. Most amphibians, such as frogs and tadpoles, have a larval stage which lives in water, even if the adults live most of their time on land. In the Devonian, amphibians grew larger and more successful, their legs and lungs adapting to their terrestrial surroundings.

One group was particularly diverse and successful. The temnospondyls were like giant salamanders, some of which measured up to 5 m (16.5 ft) in length. This group survived for 210 million years – longer than the dinosaurs were around. Over time, some temnospondyls developed hard scales on their bodies which would have helped to lock in moisture when they were away from water. None of today's amphibians has scales. They have only been retained by the reptiles that evolved from amphibians.

Eggs, hard and soft

Becoming a reptile wasn't simply a matter of evolving scales and moving away from water. Reptiles were the first amniotes: they lay eggs with an impervious shell that

don't need to be kept wet. The reptilian egg contains an amnion: a membrane which contains the growing embryo, its surrounding fluid and nutritive yolk sac. Amniotic eggs can be retained in the animal's body (as in mammals) or laid and incubated externally (as in birds and reptiles). The development of amniotic eggs was one of the most significant steps in evolution. Eggs for external incubation on land have a hard or leathery exterior with pores small enough for gases to pass through. This means that, unlike amphibian eggs, they don't need to be laid in water. Reptiles gained a massive advantage over amphibians by developing eggs that could be laid far inland, thus severing the last tie with water as anything other than a necessary part of their diet. Reptilian eggs also hatch into a smaller form of the adult animal rather than a very different-looking larval form.

Temnospondyls lounged on the river banks 330–120 million years ago.

The move from spawning in water to laying amniotic eggs on land was an advantage for the ancestors of today's reptiles. But reptiles that have returned to the sea, like these turtles, are at a disadvantage because the vulnerable hatchlings must make a perilous dash to the ocean.

THE RISE OF REPTILES

While amniotic eggs meant reptiles could spread more widely, this was not a hugely significant advantage during the Carboniferous when large areas of Earth were tropical and wet. But the next period, the Permian, saw a colder and drier climate during which reptiles could more readily exploit their egg-laying adaptation.

The supercontinent Pangea was forming, which meant there were vast, dry inland areas. Slowly, continents from the north (Euramerica) and south (Gondwana) crashed together to form the larger land mass, pushing up the central Pangean mountains. Their remnants are now evident in the Appalachians in North America, the Little Atlas Mountains of Morocco and the Highlands of Scotland.

Although these mountains were slow to form, the combining continents brought their own pre-existing mountains, creating dry areas in their shadow from early in the Permian period. On land, only the reptiles were equipped to lay their eggs wherever they saw fit. Many of the amphibians died out, and some returned to the water full-time, becoming river- or ocean-dwelling species. The land soon became dominated by large, often carnivorous reptiles which diversified to adapt to many different ecological situations. But trouble was on the horizon.

The Appalachians in North America are survivors of the central Pangean Mountains formed 480 million years ago.

The Great Dying

The mass extinction event at the end of the Permian period was the most serious catastrophe in the story of Earth. Taking place 252 million years ago, it annihilated

70 per cent of all land-going species and 96 per cent of aquatic organisms; it is appropriately known as the 'Great Dying'. It's likely that it happened in two pulses about 200,000 years apart, with some recovery in between. Vast areas of the land and oceans were left barren and deserted and food chains were torn apart, from microbes to the largest reptiles and fish. Nothing similar has ever happened before or since.

The end-Permian extinction only attracted the serious attention of paleontologists in the 1980s, after Walter and Luis Alvarez proposed that there had been a mass extinction of the dinosaurs and many other living creatures around 66 million years ago (see page 179). During the 1990s, it was posited that the reason for the end-Permian event was climate change in the form of extreme global warming. The warming was probably the result of carbon dioxide released by volcanic eruptions that formed the Siberian Traps. The eruptions spewed out enough lava to cover the USA to a depth of one kilometre and enough carbon dioxide to raise the global land temperature by about 10°C. The sea temperature increased by up to 20°C, possibly peaking at a crippling 40°C. The eruptions also removed 76 per cent of the oxygen from the oceans, so marine life suffocated. With the rise in temperature and lack of oxygen, much of the life on land was killed, including the large trees, plants, insects, smaller reptiles and even microbes.

Basalt from the Siberian Traps, originally part of a flood of 3 million cubic km (720,000 cubic miles) of lava.

It is also possible that the eruptions set fire to some of the vast Carboniferous coal deposits, releasing further carbon dioxide into the atmosphere. The heat of the eruptions probably evaporated dangerous chemicals out of the lithosphere in Siberia, releasing them into the air. Halogens (chlorine, bromine and iodine) from underground would also have spread around Earth in the air, producing acid rain and destroying the ozone layer. The discovery of high levels of nickel in rock layers dated to the Great Dying suggests that less volatile toxins could have been spread around, too.

In 2018, research scientist Jahandar Ramezani revealed from studies of rock laid down around the time of the extinction that there was evidence of heating before it, although the most severe heating came afterwards. Despite the stress of global warming, Ramezani found no pattern of increasing disappearances from the fossil record before the extinction event. The extinction had been very sudden, possibly occurring over just a few hundred years, although the final trigger is unknown.

Environmental catastrophes have tipping points, beyond which misfortune becomes cataclysm. When plants die, herbivores

MASS EXTINCTION EVENTS

DATE	CAUSE	IMPACT
444 mya, end-Ordovician	Global cooling, producing an ice age. Possibly triggered by the uplift of the Appalachians, their newly exposed silicate rock absorbing carbon dioxide from the atmosphere.	86% of species lost
375 mya	De-oxygenation of the seas, possibly caused by land plants pulling nutrients from the ground. The nutrients would then wash into the sea, causing algal bloom which used up oxygen.	75% of species lost
252 mya, end-Permian	Global warming, excessive carbon dioxide and de-oxygenation of the seas caused by volcanic eruption of the Siberian Traps.	90% of species lost
200 mya, end-Triassic	Cause unknown	80% of species lost
66 mya, end-Cretaceous	Asteroid strike	75% of species lost

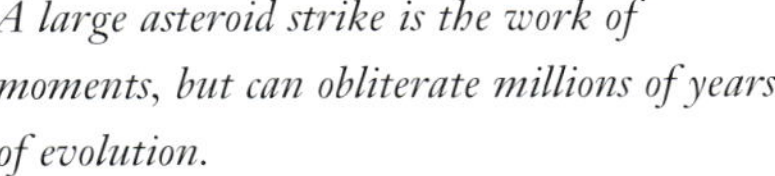

A large asteroid strike is the work of moments, but can obliterate millions of years of evolution.

soon follow, and then the carnivores that feed on them. There is a feast for the scavengers and decomposers when there's a famine for everyone else – for a while. When contamination and debris are washed into the sea and rivers, devastation spreads to aquatic environments, even if they had escaped the first impact.

Extinction is the flipside of evolution. Although Cuvier had demonstrated that some organisms go extinct, the scientific world was slow to accept even individual extinctions. Lyell argued against extinction, suggesting that although an organism might die out in one area as the result of local conditions, it could be reintroduced from another area, where in different conditions it would have survived. This might sound naïve, but given that Lyell opposed the idea of global catastrophes shaping the fossil record, it is understandable to a degree. If all catastrophes were only local, and animals were liberally sprinkled around similar environments worldwide, it should be possible to reintroduce them – as we do now. But the middle ground (and the true situation) is a combination of gradual change and catastrophic events. While the latter have not been quite as important in shaping the geological history of Earth as was once believed, they have been hugely important in dictating the path taken by life.

Wipe out

It's a large step from noticing the disappearance of a single organism to realizing that whole swathes of organisms disappeared at the same time, and then accounting for that loss. The change in the fossil record 250 million years ago is so extreme that it was spotted 150 years ago. The English geologist John Philips produced a chart of the prevalence of fossils over time and the dips in the numbers correspond well with the now-recognized extinction events.

Paleontologists generally accept that there have been five mass extinction events (see box on page 173) excluding the one which killed anoxic microbes more than two billion years ago. These five events killed at least 70 per cent of species, but this cut-off point only helps us to categorize them, it doesn't reflect any absolute difference between mass and non-mass extinction events. There have been many more episodes in which less than 70 per cent of Earth's species died off in a short period.

In geological terms, a million years is a short period of time. The extinction events now recognized may have been triggered by an event that just took a moment – an asteroid hitting the coast of Mexico one afternoon, for example – or an event sustained over a longer period than the whole of human history. The 'press/pulse' model proposed in 2006 by Nan Arens and Ian West suggests that mass extinctions generally require two types of cause to act concurrently: the first is long-term pressure on the ecosystem, such as extended volcanic activity or climate change (a 'press' stimulus); the second is a sudden catastrophic change following a considerable period of ecological pressure (a 'pulse' stimulus).

Back on four feet

It took life on Earth around ten million years to recover from the Great Dying. The

FOUR, FIVE OR SIX EXTINCTIONS?

The term 'mass extinction' was coined by Norman Newell in 1952. The five mass extinctions since the Cambrian Explosion were identified in 1982 by David Raup and Jack Sepkoski.

In 2015, the 'big five' story was disrupted with the confirmation that an extra extinction event, first reported in 1993 by Chinese paleontologist Jin Yugan, was also a mass extinction. The Capitanian extinction is said to have occurred 262 million years ago, just ten million years before the Great Dying. It was discovered by plotting the number of species of fossils disappearing from the record and spotting a massive peak. In 2020, research was published that cast doubt on the existence of the extinction event 375 million years ago.

Keratocephalus *was a herbivorous therapsid that lived in the late Permian, around the time of the proposed Capitanian extinction.*

first to do so were 'disaster taxa'. A disaster taxon is a kind of organism that populates a region during and after a local or global catastrophe lays waste to the ecosystem. Also known as 'pioneer organisms', disaster taxa are opportunistic, spreading into and exploiting unoccupied areas and flourishing as the world begins to recover from a disaster. As other organisms return or evolve, disaster taxa are squeezed back into a smaller, marginal ecological niche.

Among the disaster taxa that flourished after the end-Permian were *Lystrosaurus*, a sturdy herbivorous therapsid about the size of a pig, the marine brachiopod *Lingula*, and the seed fern *Dicroidium*. In place of the diversity that preceded the extinction, *Lystrosaurus* alone represented about 90 per cent of land-going vertebrates.

It is estimated that it took between four and 30 million years to restore ecological diversity. When recovery came, it brought some of the most iconic animals ever to tread the Earth: the dinosaurs.

Left: Lystrosaurus *thrived in the aftermath of the Permian mass extinction.*

Right: Dicroidium *was another survivor of the end-Permian extinction. Its fossils are found across South America, Australasia, South Africa and Antarctica.*

Discovering dinosaurs

It's difficult now to imagine *not* knowing about the existence of dinosaurs, but they were only recognized in the 19th century. The first known dinosaur fossil to be unearthed was described by Robert Plot in 1676 or 1677, though he didn't realize what it was. At first he thought it was a

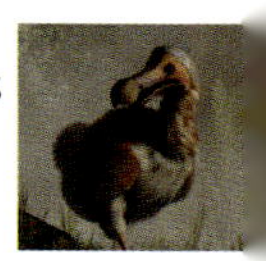

DEATH IS WORSE ON A SUPERCONTINENT

A 2015 study by Paul Wignall concluded that mass extinctions are worse on supercontinents. When Earth's land is all of a piece, the system is very bad at removing additional carbon dioxide caused by events such as volcanic eruptions or asteroid strikes, so dire global warming is easily triggered. Our current arrangement of landmasses is partially responsible for the absence of mass extinctions since the KT event 66 million years ago.

bone from some kind of elephant, then suspected it might be the end of a giant human femur. Plot didn't name it, and the fossil has long since been lost, but 90 years later Richard Brookes copied his drawing and named it *Scrotum humanum*. In theory, this name should now be used for *Megalosaurus* (the real owner of the femur) because as the first name given it takes precedence. The International Commission for Zoological Nomenclature has ruled to rescue the dinosaur from the humiliation of being named after a scrotum, so it keeps the name *Megalosaurus* which was bestowed on it by William Buckland in 1824.

The name 'dinosaur' (meaning 'terrible lizard') for a group of extinct large reptiles was coined by the English biologist Richard

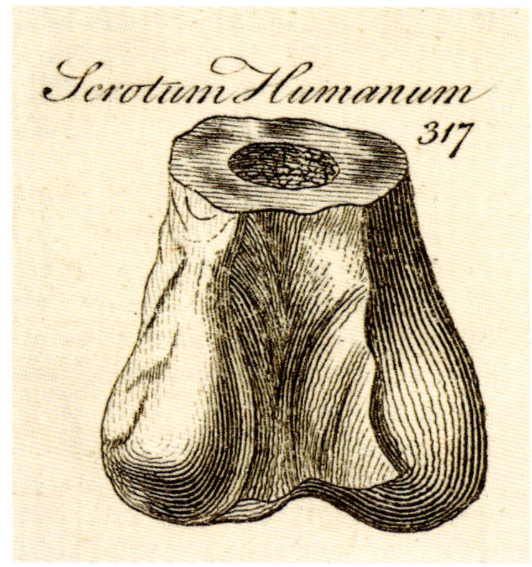

Plot's drawing of the bone he could not identify, which is now thought to be the end of a Megalosaurus *femur.*

Owen in 1842. At that point, only three dinosaurs had been uncovered, all of them in England: *Megalosaurus*, *Iguanadon* and *Hylaeosaurus*. Other large reptile fossils had been found, however. Together with her father and brother, Mary Anning had found fossils of the fish-like *Ichthyosaurus* in 1811 and the large marine reptiles called plesiosaurs in 1821. In 1808, Cuvier found a giant marine reptile later named *Mosasaurus*, and identified a flying reptile he called *Pteradactylus*.

Cuvier was the first to speculate that there had been an 'age of reptiles', and it was becoming clear that he was correct. In 1824, Gideon Mantell realized that fossil teeth he had found two years previously were those of a giant lizard-like animal he named *Iguanadon* (because the teeth were like those of an iguana). In 1831, he

published a paper on 'The Age of Reptiles' in which he suggested a geological age divided into three (to reflect the three distinct rock strata in which the fossils had been found) when many types of large reptile roamed Earth. We now recognize the age of dinosaurs as spanning the Triassic, Jurassic and Cretaceous periods.

Thick and fast

Once people started looking out for the fossils of dinosaurs and other reptiles, they appeared thick and fast, in quarries, along coastlines, on the shores and in riverbeds, in eroding cliffs and canyons – any place where rock of the right age was exposed. Germany yielded *Archaeopteryx*, a small feathered animal that looks like a transitional form between dinosaurs and birds (though it turns out that birds evolved from similar creatures in China). And very soon, North America would reveal dinosaurs bigger than anyone had ever imagined.

Just before the discovery of the first complete *Archaeopteryx* in 1861, William Foulke heard about some large bones dug up on a farm in New Jersey in 1858. He organized their extraction and pieced together the *Hadrosaurus*. In 1868, it became the first mounted dinosaur skeleton anywhere in the world. Various other fossils turned up in the east of the United States, but American dinosaur paleontology really took off in the 1860s on the west of the continent.

Land of bones

During the Jurassic period, North America was split vertically by the Western Interior Seaway. The great North American dinosaurs with which everyone is familiar, from *Stegosaurus* and *Diplodocus* to *Triceratops* and *T. rex*, all lived to the west of the sea. In 1870, paleontologist Othniel Marsh led a fossil-hunting expedition to the west. Over the coming years, he and his rival Edward Drinker Cope raced to find, extract and name as many dinosaurs and other fossils as

Astonishingly detailed fossils of Archaeopteryx *reveal its feathers.*

they could. Their 'bone wars' and personal rivalry continued until Cope's death in 1897, by which time both men were financially ruined by their dinosaur-hunting exploits. Despite their terrible behaviour during the period, Cope and Marsh unearthed a dazzling display of North American dinosaurs. In 1902, *T. rex* was added to the haul by Barnum Brown. Over the course of the 20th and 21st centuries, more and more dinosaurs have been found in new sites uncovered in South America, Africa, China, Mongolia and the Indian sub-continent. Dinosaurs have been discovered on every continent, including Antarctica, confirming Cuvier's suspicion that there really was an age when giant reptiles roamed Earth.

Triceratops *tramped across North America 68–66 million years ago.*

All good things come to an end

Today, we only have birds to represent the long extinct dinosaurs. The non-bird dinosaurs disappear from the fossil record very suddenly, around 66 million years ago. For a long time, the reason for this was a puzzle, but it was solved in 1980 by a young American geologist, Walter Alvarez – with a little help from his father Luis, a Nobel-prize winning physicist.

While in Italy studying rock formations near Gubbio, where one section 400 m (1,312 ft) deep represents 50 million years of ancient seabed, Alvarez found a thin layer of clay, just a centimetre thick, separating two layers of limestone. Both layers contained foraminifera (tiny, single-celled marine protists that build a shell around themselves). The lower layer contained large and diverse foraminifera, but the upper layer showed much less diversity and smaller individuals. Alvarez had been studying magnetic reversals and was able to date the clay layer to the end of the Cretaceous and the beginning of the Tertiary period. This layer is now known as the KT-boundary (K from *Kreide*, the German name for Cretaceous). When Alvarez realized that the clay layer coincided with the extinction of the dinosaurs, it became

Luis and Walter Alvarez investigate the K-T boundary in rocks near Gubbio, Italy.

the focus of his research. The first task was to find out the length of time it had taken for the clay layer to be deposited.

Luis suggested using a radioactive element which is deposited at a steady rate to calculate the time. They chose iridium, which arrives in small quantities in space dust all the time at a known rate, and is 10,000 times more abundant in meteorites than on Earth. To their surprise, the clay layer contained 30 times as much iridium as they were expecting. They tested other exposed sections of the K-T boundary, finding that in Copenhagen, Denmark, the level was 160 times that expected. There were also spikes in iridium in Spain and New Zealand – it was a global event.

Luis Alvarez realized the iridium must have come from space. There was no spike of plutonium, so it hadn't been caused by a nearby supernova raining debris on the planet. Then an astronomer friend, Chris McKee, suggested an asteroid strike. Calculating from the iridium content of chondrite meteorites, Luis worked out that the impactor would have been 300 billion metric tonnes and 10 km (6 miles) across. It would have made a crater 200 km (124 miles) wide and 40 km (25 miles) deep.

Asteroid Armageddon

The scenario that would follow such an asteroid strike is grim. Travelling at 25 km per second (50,000 mph), the asteroid would have struck with the impact of 100 million atomic bombs, hurling vaporized and molten rock halfway to the Moon. A giant fireball would have instantly killed everything within a few hundred kilometres. Tsunami, landslides and earthquakes would have occurred almost instantaneously. Dust in the atmosphere might have blocked out the Sun for months, killing plants and wrenching apart food chains. No land animal larger than 25 kg (55 lbs) could have survived the event.

When the Alvarez team published their theory in 1980, a lot of their colleagues were sceptical. The idea of catastrophism had been replaced with that of gradual change, so the Alvarez theory seemed a retrograde step. Walter needed more evidence – a crater, for example. But there were no known large craters from the correct period; Walter began to think that the impact must

have been in the sea. Then, in a Texas riverbed, a clue was found – deposits were discovered characteristic of a tsunami that may have taken place at the time of the K-T extinction. The tsunami would have been 100 m (328 ft) tall, far higher than any tsunami previously known. In 2004, in the Indian Ocean, the devastating tsunami which killed nearly a quarter of a million people had waves of 30 m (100 ft) high. There were also tektites in the riverbed – blobs of rock that had been molten and quickly solidified while falling through the air. This was the evidence for which the Alvarez team had been looking.

This is how the K-T asteroid impact might have looked from space, 66 million years ago.

Alan Hildebrand, a geology graduate student at the time, calculated that the tsunami must have resulted from a strike in the Gulf of Mexico or the Caribbean. Further tektites from Haiti suggested an impact about 1,000 km (621 miles) away. Hildebrand identified gravity anomalies and rounded features on the seabed. It transpired that a Mexican oil company, PEMEX, may have found the crater in 1981, close to the village of Chicxulub on the Yucatan peninsula. The crater is 180 km (112 miles) across – very close to Alvarez's predicted 200 km (124 miles); it was named as the likely site in a paper published in 1991.

Reconstruction of T. rex *witnessing the asteroid crashing into the coast of Yucatan, Mexico.*

The Final Piece?

A final piece of the puzzle seems to have appeared in 2019.

A fossil field discovered in North Dakota, USA, in 2012 is packed with what appears to be evidence of the moments just after impact. There are freshwater fish killed by the fireball, with fragments of glassy rock lodged in their gills, and there are uprooted trees. Dead trees and fish are strewn haphazardly as though blasted from the Earth only to fall down again.

The site, Tanis, is 3,000 km (1,860 miles) from Chicxulub; at the time of the impact it was near a river which drained into the sea. Researchers suggest that a megatsunami more than 100 m (328 ft) tall, triggered by the shock waves of the impact, could have hurled sediment together with freshwater and marine organisms onto the land just 13 minutes later. The sediment is 1.3 m (51 in) thick and topped with a layer of iridium. It's too early to be certain that the site is what it appears to be, but if it is, we have a unique snapshot of the worst-ever day to be alive.

The Chicxulub Crater is largely underwater off the coast of Mexico. It was formed by a large asteroid or comet colliding with Earth. The crater is 20 km (12.5 miles) deep, and its diameter, at 180 km (112 miles), is the distance from Austin to Houston, Texas.

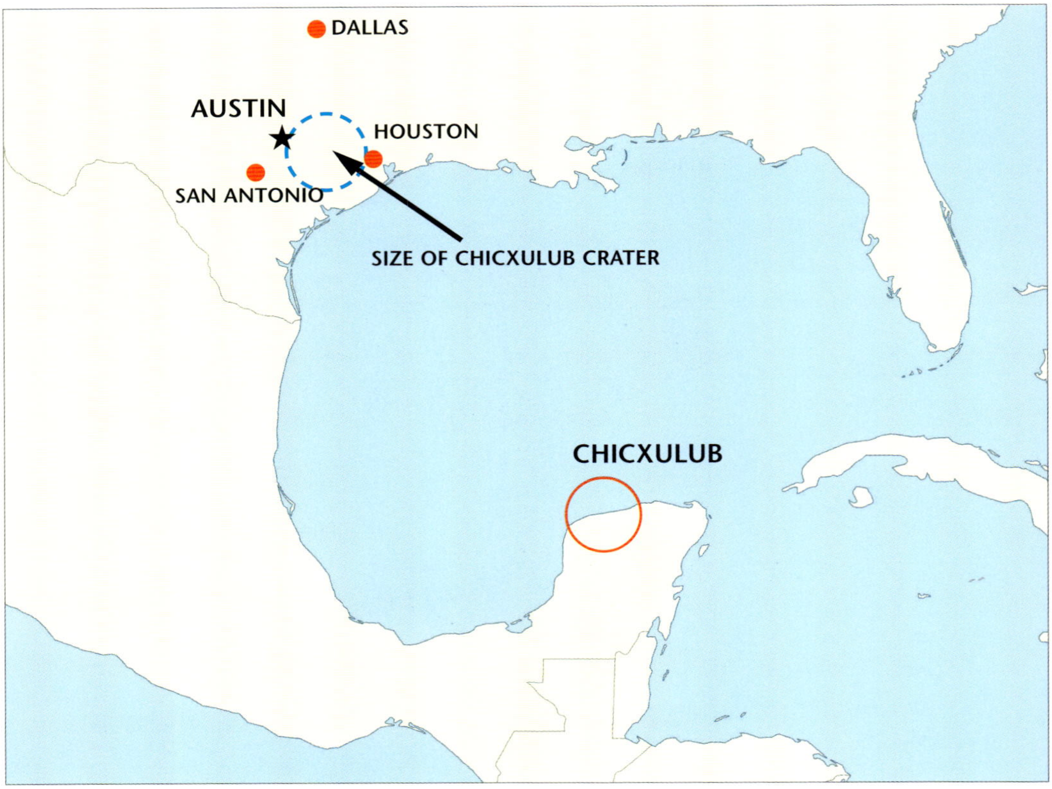

DALLAS
AUSTIN
HOUSTON
SAN ANTONIO
SIZE OF CHICXULUB CRATER
CHICXULUB

'Understanding how we decipher a great historical event written in the book of rocks may be as interesting as the event itself.'
Walter Alvarez

Evolution – Darwin and the finches

Our interpretation of the rise and demise of the dinosaurs, and of what came before and after, is heavily reliant on the evolutionary theory set out in 1859 by Charles Darwin.

In 1831, the young Darwin had just completed a geological tour of Wales when he was invited on a five-year journey as a 'gentleman passenger' on HMS *Beagle*. The voyage took him first to South America and then, famously, to the Galapagos Islands off the coast of Ecuador. Darwin was tasked with making scientific observations: in South America he studied geological features and collected fossils; on the Galapagos he collected birds, which he noted had slightly differently shaped beaks on different islands.

After his return, it took Darwin many years to write up and make known the results of his research. In 1859, he finally published *On the Origin of Species*, prompted apparently by the news that Alfred Wallace was about to publish a similar theory. In the end, papers by Darwin and Wallace were published in the same year, an elegant solution engineered by Charles Lyell.

Darwin's theory states that organisms change over time as a result of 'natural selection' – those best adapted to their environment and prevailing conditions ('fittest') are most likely to succeed, to grow and reproduce. They find the best food and living spaces and are successful in sexual competition. Consequently, their features

Darwin's finches evolved different beaks according to the food available on the islands where they lived.

are passed on to future generations and intensified over time. Less well-adapted individuals are less likely to be successful and their characteristics will be lost over time. By a process of gradual evolution, one species can change into another.

The famous example of the birds he found in the Galapagos Islands illustrates this perfectly. All the birds had a common ancestor, a species that arrived on one of the islands from the mainland. Over an extended period, as the birds spread between islands, they developed beaks adapted to the different diets the various islands afforded. Those with beaks best adapted to eating seeds were most successful on islands with plentiful seeds, for example. The birds diverged into different species, each well adapted to prevailing conditions in its own environment.

From evolution to genes

Darwin could not explain the biological mechanism of inheritance. The lack of a means by which evolution worked must have pleased his many detractors. Many people were reluctant to relinquish a belief in the uniqueness of human beings, seemingly undermined by Darwin's suggestion that we have evolved from apes. Human evolution, which Darwin did not emphasize in his book, was a focus for dissent. Other arguments, raised by Creationists, maintained that there were too many 'missing links' – organisms in between the ones we know about which must surely have existed if the theory was correct – for which we have no evidence. Few organisms have been fossilized, and of the small number of fossilized organisms that do exist, few have been found. It is not surprising, therefore, that we don't have a complete and continuous record of every type of organism ever to have lived. But gradually gaps, like Romer's Gap, began to be filled.

In the 1920s, the American biologist Thomas Hunt Morgan discovered the role of genes in heredity through his experiments of breeding fruit flies. In the 1860s, an Austrian monk, Gregor Mendel, had discovered the patterns in which characteristics are inherited, but he could not explain how features passed between generations.

Some things change, others stay the same: the fern Osmunda claytonia *has remained almost unchanged for 180 million years. The same plant has been found in the fossil record as far back as the Triassic period.*

Back to the sea: cetaceans such as dolphins evolved from land mammals colonizing the oceans.

Morgan's work explained the mechanism of evolution. A model now known as the Modern Evolutionary Synthesis emerged during the 1930s and 1940s. It was bolstered when Francis Crick, James Watson and Rosalind Franklin unravelled the molecular structure of the material of heredity, DNA, in 1953. Understanding the genetics of evolution has since enabled humankind to change its course, engineering organisms to serve specific purposes.

From dinosaurs to now

The K-T extinction was catastrophic for many organisms, but among the survivors were mammals and the last of the dinosaurs – birds. Life followed the familiar pattern of diversification by adaptive radiation – organisms expanding into newly available niches and adapting to specialize there.

Small mammals had evolved in China in the late Triassic period. They were often nocturnal and probably kept well out of the way of larger animals, living in burrows underground or in trees. However, around 10–15 million years after the K-T extinction, mammals began to grow larger and spread into all kinds of ecological niches, even returning to the sea where their legs evolved into the fins and flippers of whales and dolphins. This was the start of the Cenozoic era, which continues to this day.

Scattered lands

By the end of the Cretaceous, Pangea had broken up and the continental landmasses were moving towards their current positions. South America and North America were still separate, and both were closer to Europe

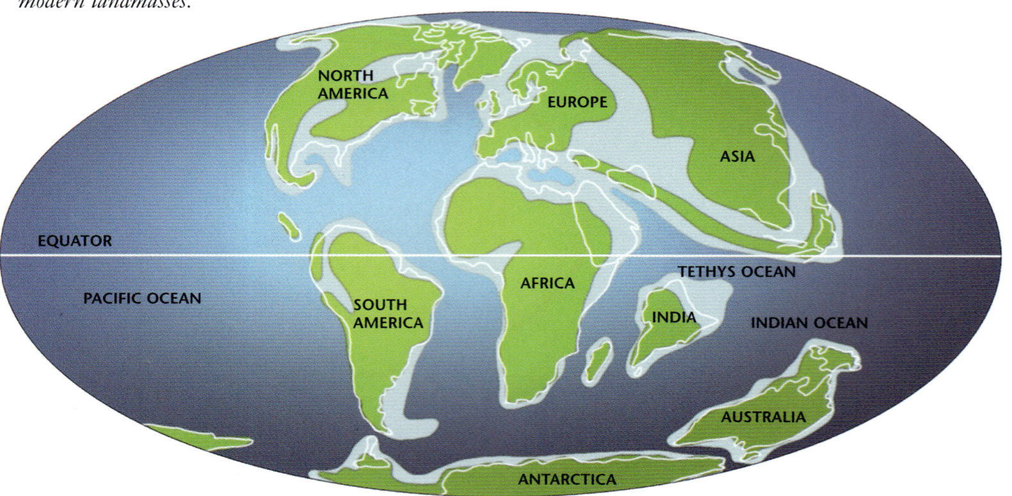

Earth as it looked around 66 million years ago, at the end of the Cretaceous. The white outlines correspond to modern landmasses.

and Africa than they are today. The Atlantic Ocean was just opening up and India was an island that was moving northwards and heading for a collision with the rest of Asia, which would push up the Himalayas. Australia was an island off the coast of Antarctica. The map was recognizable, if not quite the same as it is today.

With this new distribution of land, the coastlines became more extensive – there were relatively few areas very far from the sea. With the emerging mountain ranges, more of Earth's rocks became exposed to the elements. Carbon in the atmosphere combined with water to form a weak acid and fell to the Earth's surface as rain, where it dissolved the rocks (chemical weathering). There was a gradual reduction of carbon dioxide in the atmosphere. As the blanket of greenhouse gases thinned, the temperature dropped. Throughout the Cenozoic era the world has become progressively cooler (although there have been colder and warmer patches along the way).

Grass and grazers

As the climate changed, the humid forests retreated and the land opened up. About 25 million years ago, grasses began to colonize vast areas. With the grasslands came animals whose teeth gradually adapted to breaking down the tough, fibrous leaves and coping with the grit picked up while doing so.

It is not clear whether grasslands and grazers emerged together, or whether the grasslands came first and provided an environment into which the grazers evolved. Either way, the climate altered, triggered by geological change, and plants and animals adapted to cope with it.

With plenty of food, the grazers became larger and more diverse. Grass responds

well to cropping. It spreads from the roots, so the plant increases in size underground. Other low-growing plants cropped by the grazers didn't share this advantage, so grasses became dominant. Huge numbers of docile grazers provided a food supply for larger and larger predators. The ancestors of today's lions, wolves and bears emerged and hunted the grazers, which learned to group together in huge herds for protection.

Conditions favoured large mammals in other ways, too. Around 50 million years ago, the level of oxygen in the atmosphere rose by about 5 per cent, perhaps peaking at 23 per cent. This helped with the development of large bodies and brains, which need a lot of oxygen. The climate was warmer than it is today, with crocodiles rather than ice caps near the North Pole. Sea levels were probably 100 m (328 ft) higher than they are now.

Placental mammals

A study in 2013 found that an important biological development emerged probably somewhere in the Americas, just 400,000 years after the asteroid strike – the emergence of placental mammals. These animals nourish their unborn babies inside the mother, using a specially grown organ called the placenta, and give birth to live young. It would prove a remarkably successful biological strategy – there are now more than 5,000 species of placental mammal, ranging in size from tiny rodents to vast whales. Placental mammals spread around the world quickly, though they didn't reach Australia or South America which were further away from the other landmasses. They wouldn't arrive in South America until it developed a land bridge to North America and they didn't make it to Australia until carried there by humans.

The delivery of placental mammals to Australia was just one of the changes that humans would make to Earth, starting very early in their tenancy of the planet. Humankind has transformed Earth more than any other organism apart from the photosynthesizing cyanobacteria, which put us on a path towards an oxygenated atmosphere.

Placental mammals are now found around the world, although these wild pigs in Australia could not have crossed the ocean without human assistance.

CHAPTER 9

Into the **ANTHROPOCENE**

'Bacteria . . . have been here for three-and-a-half billion years, and without them we have no chance whatsoever of survival. Humans are something very recent, like the froth on top of a glass of beer.'

James Lovelock, 1990

The rise and potential fall of humankind is just a miniscule part of the story of Earth. We have been here for a split second of geological time and will likely be gone as quickly. But we will leave our scars, just as the first cyanobacteria left theirs.

Humankind has changed Earth's landscapes in ways unimaginable even just a few centuries ago.

Out of the woods

Homo habilis, *one of our early tool-using ancestors, lived in Africa 2.1–1.5 million years ago.*

As Earth recovered from the last great asteroid impact, birds grew large and flightless, then returned to being (mostly) small and flighted, mammals grew large and varied, and reptiles took up a less prominent position than before. Meanwhile, in the first ten million years after the K-T asteroid strike, the ancestors of the primates emerged from the forests. To begin with, they were small, squirrel-like herbivores who scampered through the trees of Europe, Asia and North America, holding on with their hands and feet.

The first 'Old World' monkeys and the first 'New World' monkeys appeared about 34 million and 30 million years ago respectively, presumably from a group that somehow found its way across the ocean, perhaps drifting on a raft of vegetation. Over a long period of cooling climate, continental separation and falling sea levels, the primates continued to move around and evolve until, eight or nine million years ago, one group in Africa split into two. One branch would evolve into gorillas, and the other into chimpanzees, bonobos and humans. The first species of Stone Age hominin, *Homo habilis*, appeared 2.8 million years ago in Africa.

Making changes

Hominins began making changes to their environment almost immediately. *Australopithecus* started using tools 2.5 million years ago. The implements steadily improved over time – tools enabled humans to be more ambitious when confronting their enemies and allowed them to bring down larger prey. While the australopithecines had a mostly plant-based diet, probably augmented with occasional scavenged eggs and pieces of meat, by the time of late *Homo erectus*, meat and fish were a much more important part of the diet. This is evident from refuse sites (bones and shells in caves,

LUCY, AND OTHERS

The earliest human ancestor is currently thought to be *Australopithecus anamensis* who lived in Africa 4.2–3.8 million years ago. The more familiar *Australopithecus afarensis*, known from the famous 'Lucy' skeleton, lived from 3.9–3 million years ago, and probably branched off from *A. anamensis*. The earliest *Homo sapiens* fossils, found in Africa, are 300,000 years old. The earliest fossils of modern humans found outside Africa date from 210,000 years ago and were discovered in Greece. Other groups have been found in China (from 125,000–90,000 years ago) and elsewhere. Modern humans outside Africa evolved from a group that migrated 60,000 years ago.

and so on). *Homo erectus* was also the first in our direct line of descent to leave Africa.

The next big step was harnessing fire. Exactly when humans achieved this is uncertain. The earliest unambiguous evidence of fire use comes from a site in China dated rather vaguely to 790,000–400,000 years ago. Fire and tool use gave humans a massive advantage over any other organism. It meant they could move into different environments without having to wait for biological evolution to equip them for different conditions. People could move north into colder regions rapidly, not just at the pace at which evolution might allow them to grow thick fur and a layer of insulating blubber. They could migrate as soon as they had killed an animal with thick fur and fashioned a warm wrap from its pelt – the work of an afternoon rather than evolution over several millennia. And if they still felt cold, they could start a fire.

Humans were (and still are) a tropical animal capable of living in temperate and

With body proportions similar to modern humans, homo erectus *used tools and fire, and cooked food.*

even cold regions because of their ability to exploit fire, tools and other animals. The only substantial concession that evolution has made to humans moving out of the tropics is to equip people in northerly regions with pale skin. With less exposure to sunlight, paler skin is still able to synthesize vitamin D, but one disadvantage is that it burns if exposed to an excess of sun. Still, it's easier to stay in the shade or deal with sunburn than to struggle with rickets and other vitamin-D deficiencies.

Early warning

Right from the start, the relatively large brains and greater dexterity of humans had a negative effect on other organisms. It's probably no coincidence that there is no megafauna left anywhere in the world but Africa, where humans started out; this is possibly because the evolution of humans on the African continent allowed the large animals to adapt to them. Despite this, by the time the first hominins left Africa, the average size of African mammals had halved. In every land to which humans migrated, the megafauna disappeared very soon after their arrival. This was probably the result of hunting: working together and using tools, humans could tackle even large prey. Climate change and habitat destruction might have played a part as well.

A group of Neanderthals, wearing clothes and carrying weapons, attack a bison. Neanderthals survived until 40,000 years ago, coexisting with Homo sapiens *in Europe.*

A study in 2018, by Felisa Smith of the University of New Mexico, quantified the change precisely. The size of mammals in every ecoregion declined rapidly once humans moved in; the mammals that died out were 100 to 1,000 times larger than those that survived. The pattern has been repeated on every continent except Antarctica (which doesn't have large animals) over at least 125,000 years. It's not just *Homo sapiens* that is responsible – the decline probably started with *Homo erectus* and other species up to 1.8 million years ago. When humans moved into Europe and Asia, the size of mammals halved, as it had in Africa. When they moved into Australia, it declined by 90 per cent. In North America, the average mass of a mammal fell from 98 kg (216 lb) to 7.7 kg (17 lb).

Volcanic winter

Humans have not had to cope with a devastating extinction event yet, but it is possible they almost did so about 75,000 years ago. Around that time, the Toba supervolcano in Indonesia erupted catastrophically, producing around 100 times as much ejected material as the eruption of Tambora in 1815, at around 3,000 cubic km (720 cubic miles). Studies of the effects of the eruption are inconclusive. Geologist Michael R. Rampino and volcanologist Stephen Self claim it caused a 'brief, dramatic cooling or "volcanic winter"', but evidence from Greenland ice cores suggests a thousand-year-long cold period. Other experts say there was moderate cooling for a short time, and some say that there was no significant effect.

Genetic studies have shown that humanity and some other species, including chimpanzees and tigers, encountered a

Lake Toba seen from space; it is the site of a supervolcano eruption around 75,000 years ago.

genetic bottleneck at around the same time. This signals a massive reduction in the gene pool caused by a large part of the population dying. The human population probably fell to around 3,000–10,000 individuals, from whom all modern humans have descended. Some scientists believe that the bottleneck and the eruption are unrelated, but there was clearly some type of crisis and humans narrowly avoided extinction.

One of humankind's earliest permanent changes to the surroundings was to use pigments to paint on the walls of cave dwellings.

Ganging together

Many organisms live in groups, but humans began to live in larger and larger groups, changing the landscape and trading with other groups of people. The Neolithic Revolution, which began about 12,500 years ago in different areas of the world, saw a change from nomadic hunter-gatherer lifestyles to communities settling in one place and farming the land.

The impact of humans upon the natural world began to escalate as agriculture brought land clearance, small-scale deforestation, the re-routing of waterways, and the establishment of irrigation. It also affected the genetic make-up of other organisms through selective breeding. Humans were the first organisms to make lasting changes to the physical landscape. Instead of simply making nests and burrows that are gone in a few years, people dug rock from the ground in one place and moved it to another, then fashioned it into shapes it would never attain naturally. They mixed clays and pigments to make pottery, and later extracted minerals and metals from the ground and separated or mixed them. The days of a truly natural world were over.

A rising tide of humanity

Settling into towns and cities might have damaged human health (see panel opposite) but it didn't prevent people from procreating. The human population boomed, and with spoken and then written language, people were able to share and pass on knowledge, to work cooperatively on projects that could last more than a lifetime, to trade objects around the world and to begin to build the edifice of modern science. They invented stories to explain the world around them, producing religions and the creative arts. They built social structures based in laws and they invented money. In short, they became modern humanity and their numbers grew and grew, as did their impact on the rest of the world.

As humans changed from a largely nomadic and peripatetic lifestyle to settled and even sedentary lives, their bodies also changed.

FROM NOMAD TO FARMER

It may have seemed a good idea for people to settle into groups, which would later become towns and cities, and practise agriculture to ensure a regular food supply. But the impact on human health was not entirely positive. Living in close quarters with others resulted in the spread of disease and the first epidemics. Close proximity to domesticated animals enabled the transfer of animal pathogens to humans, and their subsequent evolution to human pathogens. Examples of infectious diseases that spread from animals to humans include smallpox, influenza and measles. Also, the nutritional standards of settled populations tended to be worse than those of hunter-gatherers. The transition from meat-based to cereal-based diets resulted in a reduction in stature and life-expectancy. It would take until the 20th century for human height to return to the level it had been before people decided to settle.

Nevertheless, farming had its benefits. The availability of milk and cereal grains meant that mothers could raise a younger and a slightly older child concurrently, so the population increased more rapidly. The settled communities learned to store surplus food so that they could still eat in times of want.

Moving around

Humans were the first land-based species (besides birds) to cross vast expanses of open sea. They took other species into these new regions as well, sometimes on purpose and other times as inadvertent passengers. This was sometimes detrimental to other humans when, for example, diseases were carried into vulnerable populations; it could also be detrimental to other organisms and environments. Carrying rats, dogs and other predators to islands that were free of these creatures often destroyed populations of endemic animals. Other introductions, though not directly predatory, out-competed the local species and drove them to extinction.

Taking over

Humans have become immensely successful as a species over a very short time. This has given other species little chance to adapt to the changes we have brought. The human population, no more than 10,000 at the bottleneck 75,000 years ago, reached one billion around 1804. Just over 200 years later, it is now nearly eight billion. Advances in medicine and food production have freed our population from the constraints that affect other organisms. We are not limited by the supply of food we can find, because we can grow more. We are alone in being able to combat disease, and move over very long distances en masse and very quickly. The world has not had to deal with an

Zebra and quagga mussels originated in eastern Europe but were carried to North America in the bilge water of ships. They have colonized the Great Lakes, where they have so reduced the plankton density of the water that the clearer waters are prone to deadly algal bloom.

organism like us before, and the future is hard to predict.

Without seeking to minimize the earlier impact of humans, things certainly altered significantly and rapidly with the advent of the Industrial Revolution. The greatest societal change since the Neolithic agricultural revolution, the Industrial Revolution started in Britain and northern Europe and rapidly spread to North America and elsewhere. Machinery began to automate tasks previously done slowly by people, and the machinery began to be driven by fuel rather than human or animal muscle power. Steam power came first, using wood and then coal as its energy source. Coal mines opened everywhere that coal could be found, and canals were dug to move the coal and products of the new factories that the coal powered. Here our story crosses itself, for it was the digging of canals and coal mines that triggered the geological discoveries of the 18th century. Although we tend to think of industrialization as a feature of the modern age, people used fossil fuels long before the 18th century. The city of Babylon used asphalt 4,000 years ago, the Greek historian Herodotus described a well for oil and bitumen, and the Chinese extracted petroleum 2,000 years ago. The Chinese also drilled oil wells in the 4th century, connecting them to salt-water springs by bamboo pipelines, burning the oil to evaporate seawater and extract salt.

However, these small-scale operations were nothing compared to what came with the Industrial Revolution. Within 100 years, towns and cities became dirty and polluted places where factories spewed out smoke and poisoned the water and people worked in deplorable and dangerous conditions. The towns and cities grew ever larger as people moved there to find work. Farming adapted, too: machinery made up for and exacerbated the labour shortfall, and the resulting shift to larger farms and larger fields changed the landscape.

From ground to air

The Industrial Revolution set us on the path we are still following, burning fossil fuels and thereby removing stored carbon from the ground and releasing it into the air in the form of carbon dioxide. The invention of motor vehicles driven by petroleum or diesel in the late 19th century, and the discovery that natural gas could also be burned as a fuel, compounded the damage. Scientists can measure the quantity of carbon dioxide in the air in the geologically recent past by extracting bubbles of trapped atmosphere from ice cores drilled from polar regions. This has revealed that the level of carbon dioxide in the atmosphere is far higher now than it has been at any time in the past 800,000 years, and has risen rapidly in the last 200 years. There is no plausible explanation for this other than the release of carbon dioxide from burning fossil fuels.

Rising levels

The first quantitive evidence of rising carbon dioxide levels came to light in 1938 when the English engineer and amateur meteorologist Guy Callendar compared recent measurements of atmospheric carbon dioxide taken in the eastern USA with historic records from Kew in England from

1898–1901. The level at the turn of the 20th century was 274 ppm, but by 1938 it had risen to 310 ppm. Callendar concluded that the cause was emissions from burning fossil fuels. The idea had been proposed earlier, by the Swedish chemist Svante Arrhenius, but Callendar was the first to provide solid evidence of rising carbon dioxide levels.

The rate at which atmospheric carbon dioxide is increasing is recognized as cause for significant alarm by all but the most intransigent climate refuseniks. Extreme weather events, higher temperatures and melting ice around the world are signalling serious trouble ahead. But when Callendar first noted the change he felt it would save humankind from a returning ice age, which seemed more of a concern in the mid-20th century than that of global warming. In 1958, scientists began to track the level of carbon dioxide in the atmosphere above Mauna Lao, Hawaii. The graph of those readings shows the proportion relentlessly rising, but with a characteristic saw-tooth line that reflects the pattern of higher fuel use in the northern (more populated) hemisphere in the winter months. The graph is called the Keeling curve, named after David Keeling who began the programme in 1958 and directed it until 2005.

Yesterday's climate

To understand modern climate change, we need to put it into the context of earlier climates. The Neolithic Revolution came at the end of a glacial period in which ice had extended well into Europe from the North Pole. As the climate grew warmer, it became possible for people to move north and to farm reliably. The temperate climate in Europe and Asia helped humans to flourish. While rock formations and fossils have provided geologists with ample evidence with which to work out Earth's physical and biological history, climate is ephemeral and leaves fewer traces. Studies of paleoclimates (climates of the distant past) rely on proxies such as rocks damaged by ice, pollen and spores, and isotopic signatures in rocks.

Producing climate

Earth's climate is produced by the interaction of five major components: the atmosphere, hydrosphere, cryosphere (ice deposits, including ice caps and glaciers), lithosphere and biosphere. Between them, they dictate the climate and therefore the weather. But they are constrained by the amount of heat energy available on Earth and this is determined by the Sun, volcanic activity and the presence of greenhouse gases that control how much heat can escape into space.

Out of this world

The Sun was smaller and emitted less heat in the early days of the solar system, but heat from inside Earth compensated for this sufficiently to allow liquid surface water. The amount of radiation we receive from the Sun is still not fixed. It is affected by solar activity and by variations in Earth's movement in space.

In the 1920s, the Serbian geophysicist Milutin Milankovic proposed that climate is affected by variations in the cyclic pattern of Earth's movements in space. These effects are now called the Milankovitch cycles. The most striking in terms of its correspondence with changing temperatures is the

eccentricity of Earth's orbit around the Sun (how elliptical it is, or how far it deviates from being a true circle). This varies on a cycle that takes 405,000 years to complete. The eccentric orbit is caused by interaction with the gravitational pull of the other planets, especially Venus (because it's close) and Jupiter (because it's large). By studying layers of sediment in rock cores drilled from a petrified forest in Arizona, scientists have found that this cycle affecting climate change has remained unchanged for at least 215 million years.

The eccentricity of Earth's orbit interacts with other cycles to produce a pattern of varying insolation (incoming sunlight) over 100,000 years, which matches the patterns of glaciation in the last 2.58 million years. Other variations in Earth's movement in space include its axial tilt and the gradual shift of its orbit (called apsidal precession). On an even larger scale, the entire solar system is revolving around the centre of the galaxy on an orbit that takes 230 million years. The effects of these cycles on Earth's climate are still being investigated.

Drilling a core (cylinder) from rock reveals a chronological record extending over millions of years.

Back to Earth

The climate of Earth has varied far more over the last four billion years than would result just from the cycles of Earth's movement in space and the Sun's activities. The composition of the atmosphere, nature and position of the land and the activity of all kinds of organisms combine to create the climate within the parameters set by these cycles. Discovering details of Earth's earlier climate becomes increasingly difficult the further back in time we attempt to look.

CLIMATES OF THE PAST

Interest in paleoclimates began in the 19th century. In 1889–91, the New Zealand geologist John Hardcastle described the first record of previous climates based on the loess deposits of Timaru on South

Island, New Zealand. He realized that loess, a type of silt, is deposited as a layer of wind-blown dust which then compacts and hardens. It typically forms during glacial periods when there is little vegetation to prevent the removal of soil. Hardcastle saw that the deposits at Dashing Rock record several glacial periods. Between glaciation, darker rock is laid down, clearly separating the periods.

Loess deposits in New Zealand have revealed useful information about paleoclimates.

Archives and proxies

Hardcastle's work was the first example of a scientist using a proxy (sediment) to access data about the climate recorded in it. Today all but the most recent climates are studied through proxies and the archives they produce. Human records provide information about the last few hundred years, though with decreasing detail and precision as we go further back in time. For prehistory, climate data must be dug out of the biological, physical and chemical environment.

Ice cores preserve samples of ancient atmospheres in trapped bubbles, as well as dust, pollen, volcanic ash and other indicators of past climate. Current ice cores go back 800,000 years – long enough to correlate with cycles such as axial tilt and apsidal precession.

Paleoclimate researchers use three types of proxy. Biological proxies were once living: they include fossilized pollen and spores, the shells of molluscs, tree rings and charcoal. Physical proxies include the sediments studied by Hardcastle and ice cores (which are also chemical proxies). Chemical proxies include isotopes and biomarkers (chemicals produced by living things). Each of these can encode an archive of information about past climate. For example, the presence of particular types of pollen is an indication that conditions were suitable for that type of plant. Ice is laid down in layers, thick and thin layers depending on whether the snowfall was heavy or light, and these trap dust and gases which can reveal the state of the atmosphere.

FOSSILIZED RAIN

Fossilized imprints of raindrops in Prieska, South Africa, record rainfall 2.7 billion years ago. The rain fell into volcanic ash which later solidified, preserving a perfect record of a rainy day. Detailed study of the fossils reveals that the atmospheric pressure at the time was similar to or lower than it is now and that Earth was warm, with an atmosphere rich in greenhouse gases such as carbon dioxide, ethane and methane. Lyell first suggested using raindrops to calculate ancient atmospheric pressures in 1851.

Laying down our own archive

The changes that humans have wrought on the climate are also being recorded in the natural environment. The change has been so rapid that in millions of years' time it will be difficult to unpick what it records.

By the late 18th century, when George Stubbs painted The Lincolnshire Ox, *people had been refining livestock through careful breeding for millennia.*

If the ice caps melt, there will be no ice cores as an archive. Sediments and the rapid disappearance of species will show an extinction event, but the very sudden rise of carbon dioxide in recent centuries will likely be a puzzle to any future species capable of investigating.

Moulding the environment

Since the beginnings of agriculture, humans have shaped plants and animals to their needs. People have kept seed from their best crops to grow the same the next year and have bred their best sheep, cows, dogs and other animals to produce more with the traits they favour.

In the 20th and 21st centuries, we have developed the technology to control genes directly. In addition to changing the nature of plants and animals we want, we have tried to eradicate those we don't want. Often

ANOTHER MASS EXTINCTION?

The current rate of species extinction has led scientists to the conclusion that we are in the midst of the sixth (or seventh) mass extinction, this time caused by human action and its consequences: climate change, deforestation, pollution and desertification. The United Nations reports that one million species face extinction. The current rate is reported to be 1,000 times the normal rate between extinction events.

these have turned out to be – or evolved to be – quite resilient, and other unintended targets have suffered.

Following World War II, when it was used to control malaria and typhus, dichlorodiphenyltrichloroethane (DDT) was promoted by government and industry to combat agricultural and household pests. Its extensive and indiscriminate use, together with that of other pesticides, led to the deaths not only of insects but also of animals further up the food chain, and led to ecological disaster. The American marine biologist Rachel Carson documented this disaster in her ground-breaking book *Silent Spring* in 1962. Her work prompted new legislation and the start of the ecological movement.

As part of the Great Leap Forward of 1958–62, China encouraged its huge population to wage war on the 'four pests'. One of these was the sparrow, which ate grain needed to feed the population. But the intervention led to an increase in the number of insects that sparrows would have eaten and consequently to a famine that killed 15–45 million people.

We have become more aware of the fragility of food chains and ecosystems since these catastrophes of the 20th century.

Making and breaking

Humankind began extracting metals from ore thousands of years ago, enabling first the Bronze Age and then the Iron Age. We have since made alloys by combining metals in ways in which they would never mix in their natural state and we have concocted other chemicals and materials unknown in nature. Some, such as plastic and concrete, have no natural ways of breaking down and will endure for thousands of years. The radioactive waste from our power stations will also last for thousands of years.

But Earth does not stand still around us and evolution is perhaps already catching up with some of the changes we have wrought. A bacterium identified in 2016, *Ideonella sakaiensis*, can break down PET plastics using the carbon as its source of food. Another bacterium, *Deinococcus radiodurans*, discovered in 1956, can withstand extremely high doses of radiation and has been used experimentally to clean soil contaminated with both heavy metals and radioactive waste. Its unique hardiness has led to its being nicknamed 'Conan the Bacterium'.

Pesticides and disease threaten the health of bees and other pollinators upon which our food system relies.

Conclusion
EARTH, A WORK IN PROGRESS

Evolution is not over, nor is the story of Earth. Assuming that humanity survives, humans will continue to evolve and Earth will endure more hot periods, ice ages, asteroid strikes and catastrophic volcanic eruptions. The planet is middle-aged and has another five billion years or so to go before the Sun expands to devour its orbit. A lot will happen in that time. In the relatively short term, the continents will drift back together into a supercontinent; the climate will warm and sea levels will rise; greenery will reassert itself and take some of the carbon dioxide back out of the air and cool it all again. It will restore a balance that may not favour humans, but will allow something else to thrive.

Other planets in our solar system might once have supported life, and some of the moons around the gas giants could possibly

EVER-CHANGING EARTH

The carbon dioxide content of the atmosphere today is about 415 ppm, the highest it has been since the Pliocene, 5.3–2.6 million years ago. At that time, Antarctica was covered with lush forests. It wasn't the first time Antarctica had been hot; 100 million years ago, the temperature there was about the same as it is in South Africa today. It's possible that average temperatures near the tropics were 40–50 °C – today's plants and animals would not be able to survive in that heat. Alternatively, some mechanism, possibly frequent hurricanes and different ocean currents, might have redistributed heat to the poles so that the tropics were not quite as hot as it might appear. Earth has not had polar ice caps for most of its history, so if the planet is returning to a warmer state it will be nothing new.

still host life of some kind. Earth is special in having had multiple and varied life forms in many different environments, from scorching sea vents to freezing mountains, parched deserts and tropical swamps. Its ability to support life is partly accounted for by its position – close enough to the Sun for liquid water, but not so close that it is roasting. But position alone is not enough.

Earth has a unique blend of conditions, not least its tectonic activity, which makes it amenable to life. The rock cycle provides a way of refreshing the climate, jump-starting the planet out of a deep freeze by adding carbon dioxide when needed, and taking it back out of the atmosphere when it gets too hot. While Earth still has moving tectonic plates and liquid water, it is likely to remain inhabited in one or way or another. And life will in its turn continue to shape, and be shaped by, its home.

Different climates have favoured different types of organisms throughout Earth's history. As Earth warms now, who knows what type of organisms will emerge to take advantage of the new conditions?

INDEX